Industrial Pharmacy - II

Industrial Pharmacy - II

D. K. Tripathi *M.Pharm, PhD.*
Director, Pharmacy
Santosh Rungta Group of Institutions
Bhilai, CG

Ayushmaan Roy *M. Pharm.*
Assistant Professor and Head of the Examination,
Rungta College of Pharmaceutical Sciences,
Bhilai, Chhattisgarh, India

Manindra Mahapatra
Assistant Professor of Pharmaceutics
at Rungta College of Pharmaceutical Science and Research (RCPSR)
Bhilai, Chhattisgarh, India

PharmaMed Press
An imprint of BSP Books pvt. Ltd
4-4-309/316, Giriraj Lane,
Sultan Bazar, Hyderabad - 500 095.

Industrial Pharmacy - II
by **D. K. Tripathi, Ayushmaan Roy and Manindra Mahapatra**

© 2023, *by Publisher*

Disclaimer: The authors and the publishers have taken due care to provide the authentic, reliable and up to date information related to the subject. However, neither the authors nor the publisher shall be responsible for any liability for any damage caused as a result of use of this book. The respective user must check the accuracy from other sources too.

Published by:

PharmaMed Press

An imprint of BSP Books Pvt. Ltd.

4-4-309/316, Giriraj Lane, Sultan Bazar, Hyderabad - 500 095.

Phone: 040-23445688; Fax: 91+40-23445611

e-mail: info@pharmamedpress.com

www.pharmamedpress.com/pharmamedpress.net

ISBN: 978-93-95039-25-3 (Hardback)

Preface

The book, Industrial Pharmacy- II, has been prepared as per the syllabus framed by the Pharmacy Council of India for B. Pharmacy 7th semester. The syllabus has been divided into 5 chapters covering the PCI syllabus. The book has been prepared in simple language so that the students can understand the concept and contents properly. Each topic has been clearly explained. However, certain topics are excluded from the syllabus but useful and relevant to the topic. The author realizes that without these topics, it would be difficult for the readers to understand the topic clearly.

The authors have taken necessary measures to make the book error free, still the authors would remain thankful to one if he or she finds any mistake and inform through mail. The mistake highlighted will be rectified with due acknowledgement.

The course would make all the students understand at least the following:

- Know the process of pilot planting and the scale of pharmaceutical dosage forms

- Understand the process of technology transfer from lab scale to commercial batch

- Know different Laws and Acts that regulate pharmaceutical industry

- Understand the approval process and regulatory requirements for drug products

D. K. Tripathi

Acknowledgement

The authors express their thankfulness to the Management of the Santosh Rungta Group of Institutions, Bhilai and Raipur; particularly to the Chairman, Mr. Santosh Rungta whom we consider the Mentor of the mentors and a major source of inspiration, whose constant inspiration had yielded the outcome. The author expresses his sincere gratefulness to Dr. Saurav Rungta, Director (Technical) and Mr. Sonal Rungta, Director (F & A).

The authors are thankful to their colleagues for their moral support. At the last the author D. K. Tripathi remains grateful to Mrs. Bimala Tripathi who has sincerely cooperated in preparing the manuscript and without her cooperation it would be impossible to complete the task.

D. K. Tripathi

Contents

CHAPTER 1

Pilot Plant Scale up Techniques

- ➢ General Considerations
- ➢ Significance of Personnel Requirements,
- ➢ Space Requirements,
- ➢ Raw Materials,
- ➢ Pilot Plant Scale up Considerations for Solids, Liquid Orals, Semi Solids
- ➢ Relevant Documentation,
- ➢ SUPAC Guidelines,
- ➢ Introduction to Platform Technology

General Considerations

Once a new drug molecule is discovered and it passes the pharmacological and toxicological tests, it goes to the formulation development department to create a suitable dosage form. The dosage form must be stable, able to release the drug at the right place and at the right time; so that the drug can exert its therapeutic effect. The dosage form should preserve the drug unchanged during its shelf life. To select and develop an appropriate dosage form, complete information about its absorption, distribution, metabolism and elimination of the drug is to be clearly understood by the formulators.

- *Absorption a drug depends on its solubility, partition coefficient, and permeability.* Knowledge of the site of absorption is also important in choosing the dosage form.

- *Distribution* of the drug within the body compartment depends on its solubility and lipophilicity. The biological half-life of a drug depends on its protein binding property. If the drug is highly bound, its half-life will be high because the bound drug cannot permeate to body tissues and thus, its metabolism and elimination will be delayed.

- *Metabolism* is the conversion of a drug from its lipid-soluble form to a mostly water-soluble form; so that it can be eliminated from the body. The liver is the main organ in which most of the drugs are metabolized. There are some enzymes, which can metabolize a drug by chemical means – oxidation, reduction, and hydrolysis. Cytochrome P450 plays an important role in the enzymatic metabolism of a drug.

- *Excretion* is the elimination of a drug either in metabolized form or as such from the body through urine, faces, sweat, etc. The kidney plays an important role in the excretion of a drug.

A formulation should be developed in such a way that it can be manufactured economically on a large scale repeatedly without any change in its characteristic property. The formulation is initially manufactured in laboratory scale using laboratory-sized commercial equipment's and machinery. This is done to avoid the variation in quality between the large scale/commercial production due to the changes in equipment's. The lot size of the laboratory scale is also less than that of the pilot scale production. This is done to reduce the expenditure and to increase the number of lots. In both cases, the cGMP must be followed.

Once the formulation manufactured in the laboratory is found to be of consistent quality as per expectation, the formula is tried in pilot scale. In pilot scale, the lot size is increased and the method of manufacture, specifications of the raw materials used and those of the products are noted. The Fig. 1.1 expresses how the technology is transferred. All the processes are carried out as per the documented method and keenly observed. Any deviation in the method or characteristic properties of the intermediate or of the final product must be recorded. During the pilot plant scale production following points are to be documented.

- The availability and quality of each raw material
- Time required to complete each and all the processes
- Quality of intermediate at each step
- Processing parameters
- Environmental conditions for each process required
- Equipment's and machineries required and their specifications
- Space required
- Types of personnel required
- Processing loss
- Yield
- Methods of evaluation of intermediates and final product
- Specifications of the raw materials and finished products
- Type and quality of packaging material required
- Storage conditions for intermediates and final product
- Shelf life of the products

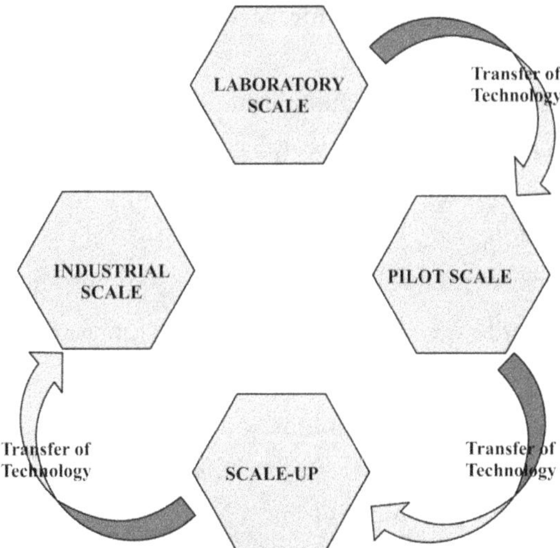

Fig. 1.1 Transfer of technology

Once the product produced complies strictly and consistently with the specifications, it may be considered that the pilot scale production is completed and the technology is ready to scale-up before transfer to commercial production. After successful completion of scale-up trials, the technology is transferred with all relevant documents to the production department for large-scale production. Thus, a pilot plant is a small industrial system, which is very similar to the large-scale production and is operated to generate information about the behavior of the system for use in the design of larger facilities. A pilot plant is an intermediate step in between the laboratory scale and industrial/large scale. These are usually smaller than full-scale production plants, but are built in a range of prototype sizes. (A pilot plant is a small-scale replica of the full-scale to provide design data for the ultimate large-scale production).

Usually, in the formulation development or research & development section, if the lot size is 1000 units, in the pilot scale, it becomes 10 000 and in the scale up, the lot size is increased to about 1 00 000, while in commercial production, the lot size becomes 5 – 10 lakhs. In other words, from R & D to the commercial scale, the lot size is increased as

$$1X \rightarrow 10X \rightarrow 100X \rightarrow (500 - 1000)\ X.$$

Objective of scale-up technique

- To develop a physically and chemically stable, therapeutically effective formulation, by optimizing various parameters.
- To create guidelines for production and process control.
- To develop specifications for raw materials handling and its requirements.
- To identify the critical steps and parameters involved in each process.
- To develop a master formula for manufacturing the dosage form.
- A pilot plant study was conducted to develop a formula identical to a commercial batch.
- Selection of infrastructural requirements to scale up the pilot plant.
- To evaluate, validate and finalize the production and processes.
- To evaluate and validate the developed product.
- To enlist the processing equipment.
- To establish the physical and mechanical compatibility of the equipment with the formulation.
- To determine the time and cost factor.

- To meet the needs of current market strategies.
- To transfer the technology from small scale to large scale production shown in Fig. 1.2.

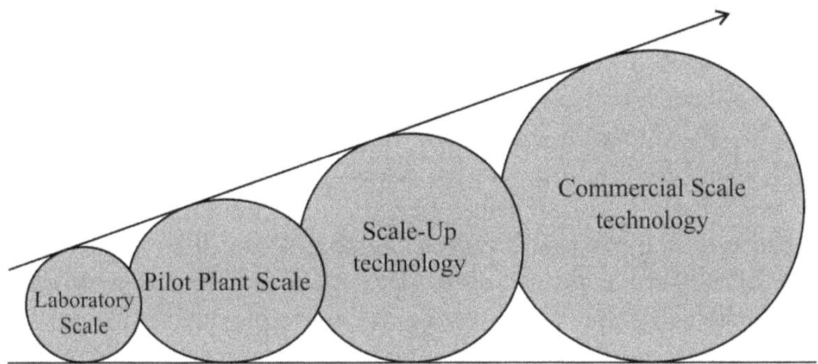

Fig. 1.2 Transfer of technology from a laboratory scale to commercial scale manufacturing

Steps involved in scaling-up of a technology

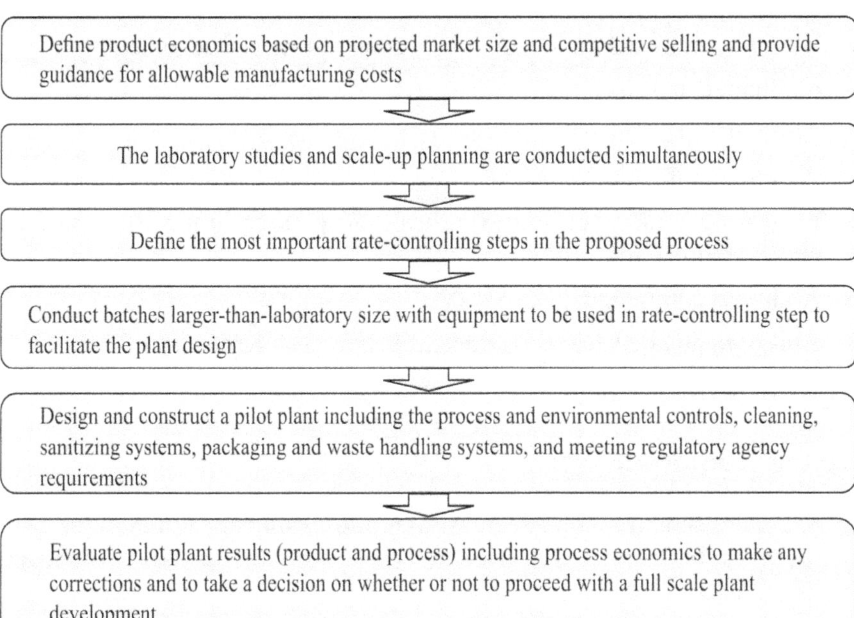

Fig. 1.3 Steps involved in scaling-up of a technology

During transfer of a technology from pilot scale and to large-scale manufacturing through scaling-up of technology following processes are followed as shown in Fig. 1.3.

Significance of Personnel Requirements

Most of the departments require people of different categories in terms of their level of education and work experience. Accordingly, they are assigned to different administrative levels. The activities in a pilot plant mostly require skill workmanship because a particular work needs to be completed within a minimum period. Persons lower than supervisor level working in a pharmaceutical pilot plant must understand why and how they are responsible and what may be the consequence of their failure. As such number of employees in a pilot plant department is limited, and depends on the number of products being developed. Usually 2–3 persons including product development scientist per product or two products are provided – an experienced scientist plus a knowledgeable and experienced operator, supported by an attendant. Therefore, the persons must have certain qualities, as listed below:

- Ability to read and understand the written directives/documents
- Intelligent analysis
- Ability to write the observations, requirements, etc.
- Possession of ethical values
- Possessing some hands-on experience of working in a pilot plant of a pharmaceutical industry
- Ability to communicate
- Ability to handle various equipment's and machineries
- Some sense of engineering.

Space Requirements

In fact, there should be four sections of a pilot plant depending on the type of work to be carried out. Hence, there should four types of space requirements –

1. Administrative and information section
2. Physical testing section
3. Pilot plant Equipment section
4. Storage section

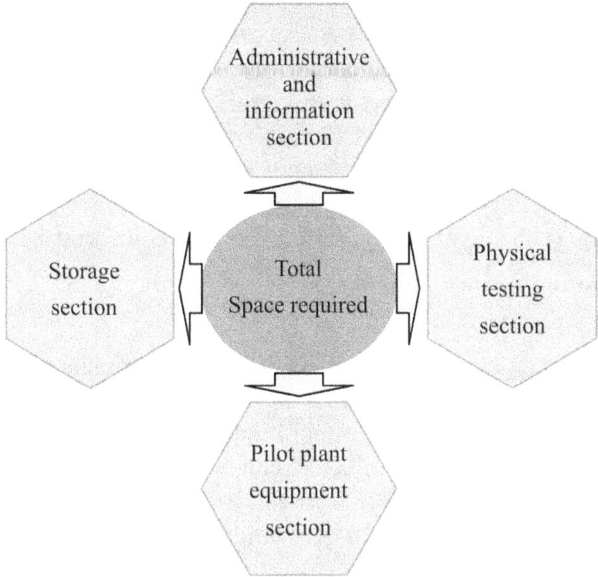

Fig. 1.4 Categorization of space required for pharmaceutical manufacturing

1. Administrative and information section

Documentation is a primary and essential activity in good manufacturing practices. Similarly, when some activities are going on information must be collected from various sources and needs to be communicated to some departments or some authorities. Thus, sufficient space should be provided to scientists and technicians working in a pilot plant to perform these activities. This space or section is called an administrative and information section. The location of this area should be adjacent to the work area but should be isolated from the work area so that the people can work without any disturbance.

The function of a pilot plant is to convert and transform the outcome of research and development activities into a feasible input of commercial production. People from the R&D department as well as those from the production department are expected to come to the pilot plant to discuss different issues. Thus, an adequate space is required so that at least five to six people can seat comfortably and discuss without disturbing other activities. At the time there should be an information centre attached to this section; where in a computer and printer must be available with internet facility to ensure that the information can be collected and documentation can be made conveniently.

2. Physical testing section

The next area or space must conduct experiments such as physical tests on the products prepared. This is a working area. There should be bench-top to keep the instruments required to perform mostly the physical tests and the samples. Commonly, the balance, pH meter, moisture balance, viscometer, desiccators, etc. Are to be placed on the bench-top. The space provided should be adequate to perform the tests comfortably.

3. Pilot plant equipment section

This is a standard pilot plant equipment floor area where the equipment required to manufacture different types of dosage forms is placed. Usually the equipment used in a pilot plant is similar to that used in an actual production department. These should be of different sizes. With the quality of scale up data collected can be assured, particularly when expensive materials are used. To evaluate the effect of the scale of research formulation and process, intermediate-sized and full-scale production equipment's would be required. In other words, to obtain the equivalence between the qualities of the products produced in a pilot plant and those produced in large scale, similar equipment's must be used in pilot scale production. This area should be used most efficiently; hence, it should be subdivided into different areas required as per the type of dosage form such as solid, semi solid, liquid, and aerosol manufactured.

Since, the dosage forms are irregularly manufactured in pilot plants as per requirement, use of pilot plant equipment's is also irregular; hence, the equipment in the pilot plant should be portable. The equipment is used only when a product is developed in the R&D department and needs to be scaled up. Thus, all the equipment is not used daily in the pilot plant department. These should be stored in a small area and be taken out before use. With this system, congestion can be removed, working space is increased and people can work comfortably. If sufficient space is there around each of the equipment's, the equipment can be cleaned easily and properly. There are equipment's which are used after cleaning in place and some are used when placed in clean area.

4. Storage section

This is the fourth area. Usually this space is found to be insufficient for storing of equipment's. Apart from equipment's, excipients, including actives (drug substances), packaging materials and finished products are to be stored orderly. All these materials should not be stored in a single area. Each category should be kept separately. Hence, the storage area should be subdivided into at least three areas – one for storing the equipment's, in the second area excipients, and in the third area-finished products should be stored. The second and third areas need to be subdivided further to reject, under test, and approved materials as per GMP. According to GMP, the packaging materials are to be stored separately, not with raw materials. The raw materials and finished products should be stored at different conditions of temperature and humidity. For example, some materials require low temperatures and humidity. Some require a dry place.

The packaging materials are usually bulky in nature and require more space. Finished products are stored as retained samples to use as a reference till it is expired. The retained samples (finished products) should be stored as per the specifications provided in the label. These samples were tested to assess their stability under different environmental conditions. Therefore, the storage area should be provided according to the requirement with different environmental conditions. For example, capsule shells require low humidity and temperature. Bottles, vials and ampoules are stored under normal atmospheric conditions. Boxes of different categories are stored under normal atmospheric conditions.

Raw Materials

The responsibility of the pilot plant scale up department is to approve and validate the raw materials required for the product. Raw materials also include an active ingredient. Because the physical characteristics of raw materials used in small pilot batches may differ when these are used in large amounts. The analytical specifications of the materials including drug substance do not change, but their physical specifications such morphology, particle size, particle shape, colour, bulk density, static charge, flow properties, the rate of solubilization, etc. It is also necessary to verify the quality of the final product by using excipients including drug substance manufactured by other companies or supplied by different vendors; because to maintain an uninterrupted production schedule there should always be some alternative suppliers or manufacturers who can fulfill the requirement of materials in terms of quality and quantity.

Processing Equipments

The attributes of final products in most cases depend on the type of equipment used in processing. The equipment should be simple in operation, economical and should be easily cleaned and maintained. In the formulation development section, equipments of small capacity are generally used. Sometimes, the equipments used in formulation development work does not match with what is actually used in large-scale production. Hence, the feasibility of the manufacturing processes is developed in formulation development and processing characteristics are determined. Once the process parameters and feasibility are established, pilot plant trials are conducted using large - scale production equipments with small capacities. If the desired equipment's is not available in house, the trial experiments of the pilot plant can be conducted at the equipment vendor's house to examine the feasibility and effect of equipment on the final quality of the product. Sometimes the quality of the intermediate product and time required to complete the process depends on the type of equipment used. For example, there are different types of mixers used in wet granulation, each has a different mixing efficiency. Hence, which mixer would be suitable for a particular product that needs to be decided? Similarly, for drying wet granules or even powders fluid bed dryer or hotair oven (tray dryer) can be used. But it may be necessary to decide the suitability of the dryer. If the capacity of the equipment is very small, it may be difficult to scale up the technology to a commercial scale. However, if the capacity of the equipment is large, the size of the experimental lots would be high and the cost of the equipment's would be high. Thus, overall expenses would be high. In fact, it would be very much difficult to run a trial experiment on an expensive drug.

When the process or technology developed is found to be realistic intermediate-sized batches are run before large-scale production.

Production Rate

In any industry, it is necessary to know how much and how long a product can be sold in the market. Depending on market requirements, the production rate should be fixed. To meet higher requirements, the rate of production should be increased and for this, equipment's of higher capacity would be required. This requires more capital investment. The following points should be considered before purchase of equipment;

- Cost of the equipment
- Whether the equipment can be used to manufacture other products or not?
- How frequently the equipment is used?
- Whether the operation and cleaning of the equipment are easy?
- Whether special training is required for the operation and maintenance of the equipment?
- What would be process-loss?

If the purchase of the equipment with a higher capacity is impossible, continuous operation of processes using equipment's of lower capacity and finally the intermediates are mixed and made a single batch. Accordingly, the production rate can be increased.

Process Evaluation

The knowledge of the effects of various process parameters as few mentioned above forms the basis for process optimization and validation. Any manufacturing process is necessary to be evaluated on the basis of certain critical parameters, as shown in Fig. 1.5.

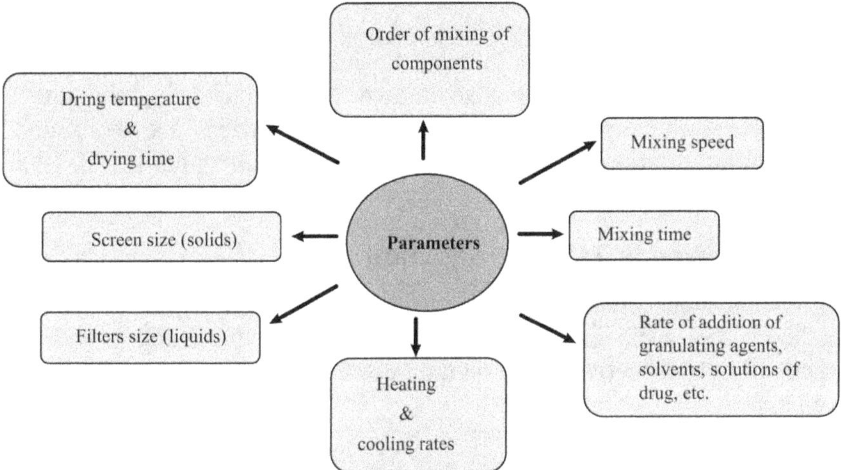

Fig. 1.5 Evaluation of the process

Pilot Plant Scale-up Considerations for Solids, Liquid Orals, Semi Solids

The primary responsibility of the pilot plant staff is to demonstrate that the product (dosage form) developed is therapeutically efficient, economical, and consistently reproducible on the production scale. The design and construction of the pharmaceutical pilot plant for every dosage form development should be such that the flow of materials, processes, and personnel are as per the norms of GMP and facilitates the maintenance and cleanliness of the area and equipment's. Irrespective of the type of dosage form, the processes are performed systematically. Each stage/step must be monitored very carefully and thoroughly starting from collection and storage of excipients to the manufacturing processes and quality of product. At each step, the tests (IPQC tests) to be performed on the intermediates are enlisted to control and ensure the desired quality of the final product. It is to be established that the the same process, same equipment's of higher capacity when used with increased amounts of material does not change the quality of the product.

Solid Dosage Form (Tablet)

When a particular dosage form is developed by the R&D department and after that is sent for pilot plant scale up studies, the formula developed is standardized. The processing equipment's to be required for large-scale production are reviewed. The rate of is very much important for any manufacturing business; through a pilot plant study the rate of production is optimized and controlled. In fact, the infrastructural requirements for commercial manufacture are also known. While increasing the rate of production (scale up activity) the critical parameters are determined. Simultaneously, appropriate records to be maintained as per the GMP are known.

Steps involved in Tablet Production

The actual steps involved in the manufacture of tablets depend on the method used such as direct compression, dry granulation, and wet granulation. Irrespective of the method used, there are some common steps such as;

- Material handling
- Granulation
- Drying
- Reduction in particle size
- Blending
- Compression

In case of direct compression, the powders are mixed, lubricated and compressed. In case of dry granulation method, the blended powders are compressed to make slugs. The slugs are then crushed to make granules; the granules formed are blended with lubricant and diluent, if necessary and then compressed. When the tablets are prepared by wet granulation method, the powders are mixed first, then granulated with paste or solution of suitable binder, passed the wet mass through appropriate sieve, the wet granules are dried, crushed to form appropriate sized granules, the granules are then blended with lubricant; if necessary, and then compressed.

Material Handling System

In laboratory operations, the materials are handled simply by hand such as scooping, drawing or pouring. In intermediate- or large-scale operations, the materials are often handled by a suitable mechanical system such as vacuum loading systems, metering pumps, screw feed system. Note that when materials are transferred for the manufacture of more than one product, there is no cross contamination. The material handling system must deliver the accurate amount of the ingredient to the formulation. The Selection of the type of mechanical handling system depends on the materials such as density and static change.

Weighing of materials is an important activity in manufacturing either in small scale or large scale. The accuracy of the strength and quality of the product manufactured depends on the addition of all materials that are accurately weighed and processed properly. Weighing in a common processing area may cause problems such as

➢ Cross contamination, and

➢ Products may become misbranded due to the presence of incorrect amount of drug or due to the presence of minute amount of drug other than the declared one.

For this reason, some industries have weighing sections that are separated from other sections. This section means only weighing activity. After each weighing, it would be checked by another person and countersigned. According to product and its lot size, the amount of drug required is calculated and then weighed. This ensures that the correct quantity of drug is used in the manufacture of each lot. Each weighing balance is regularly maintained and calibrated by an authorized agency. This is a central department responsible for correct weighing. The sanitation, particulate matter in the circulating air are controlled in the area up to the highest degree.

A separate room or area is provided to weigh a highly potent drug such as steroid or alkaloid. The area is equipped with an efficient air filter, so that even minute contaminations do not occur. However, such an area can be used to weigh the dyes.

Dry Blending

The blending of powders should be done properly, otherwise uniform mixing would not be there. That is, samples of different portions of the mixture would show different potency. To ensure uniform mixing all the ingredients should be free from lumps and agglomerates. Hence, screening and/or milling of the ingredients are done before mixing to make the mixture homogeneous and reproducible. The equipment used for blending are V- blender, double cone blender, Ribbon blender, slant cone blender Bin blender, orbiting screw blenders, Vertical and Horizontal high intensity mixers. The process of blending or mixing can be optimized by controlling the following parameters:

- Time of blending

- Blender loading

- Size of blender

It is well-known fact that the mixing of powders to make a homogeneous mixture is a challenging work. There are certain ways to achieve homogeneity. The order of addition of components should be monitored. The smallest amount is mixed with the small one; the mixture is then mixed gradually with portions of the powder of large quantity; like this. By this method, the drug can be uniformly distributed within the powder mix and variation in drug content can be reduced greatly, particularly when a small amount of drug is present in a unit dose such as per tablet or capsule. Mixing or blending of powders is generally done in a large container, which is then granulated in a mixer of similar capacity. If a manufacturer does not have such a large mixer for granulation, the powder mix (powder blend) is divided into portions and each portion is subjected to wet granulation. The dried lumps are converted into appropriate granules by milling or screening. The dried granules thus prepared are blended in a large container and the blended granules are then compressed. The process becomes more reliable and reproducible by milling or screening the lumps.

Granulation

As such homogeneous mixing of powders is a difficult task. Usually it is heterogeneous in nature and separation of particles according to their size is a common phenomenon, if particles of different size ranges are mixed. For this reason, the materials are passed through a particular sieve or mesh before mixing. The powder mixture is converted into granules of appropriate size range. Granulation is done either by dry or wet granulation method for the following reasons:

➤ After blending with suitable lubricant, the granules acquire good flow property, can flow uniformly from the hopper, can be easily compressed into tablet or encapsulated into capsules of suitable size, and uniformity of weight can be maintained.

- ➢ The bulk density of powders can be increased by granulating the powders; this helps encapsulating the granules in smaller size capsule shell.
- ➢ Granulation of the powders changes the particle size and hence, size distribution. Inter-particle binding properties during compaction or compression can be improved by changing the particle size distribution.
- ➢ Tablets of very small amounts of drug (mcg) contain higher amounts of excipient; in the granulation of a solution of drug when mixed with binder solution or paste and added to the powder, it results in a uniform distribution of the drug within the bulk powders.

Usually, Sigma blade mixer, Tumble blender, Heavy-duty planetary mixer is used for granulation. This equipment's can process 100 to 200-kg powders. Since this equipment is attached to a heavy duty motor, it can generate a high shear force. The high shear force generated by these equipment's and the weight of the material can granulate the powders within less time compared to that required during granulation in laboratory scale granulation. Such high shear mixer can increase the density of light powders.

Recently a multifunctional closed processor has been developed which can perform mixing of the powders, wet granulation, drying, sizing, and then lubrication in a continuous process. This type of equipment reduces the material handling and hence, reduces the processing loss and worker requirement considerably.

Binders are used in making tablets and it is added for granulation. In case of dry granulation, dry binders are used and for wet granulation either solution or paste of binder is used. The powders with adequate compressibility or compatibility can be compressed directly after lubrication. Otherwise, granulation must produce tablets. The properties of granules and the strength, friability, disintegration of tablets produced depend on the amount of binder used. Hence, it is necessary to optimize the amount of binder to be added per tablet or per lot of tablets. Moreover, different binders have different binding properties with respect to the concentration used; hence, the binder, its concentration and amount of solution or paste are to be optimized with respect to the properties of tablets. In some cases, the rate and time of mixing affect the characteristics of the granules. Thus, following parameters are to be standardized or optimized for effective granulation:

- ➢ Binder
- ➢ Concentration of binder

➢ Amount of binder solution or paste

➢ Duration of mixing

➢ Rate of mixing

➢ Type of mixer for granulation

In some cases, binding agent when is dissolved or dispersed in a granulating fluid or solvent, the consistency of the solution or dispersion becomes very high. Its addition to the powder mixture becomes difficult by pumping or pouring. In fact, during the pilot plant study such type of granulation is difficult to be standardized, particularly if a closed system is used. In such cases, by adding more of the solvent, the granulating solution/dispersion can be diluted; so that it can be handled easily.

If any component of the formulation, may be the drug or any excipient, whose quantity is very less, can be easily and uniformly distributed within the bulk by wet granulation. This is another advantage of this method. For example, the preservative or the drug must be uniformly present in the formulation and uniform mixing in the powder state is really difficult. By using either a solution or dispersion of these in granulating fluid, uniform mixing can be satisfactorily done. Sometimes, a mixture of water and water miscible volatile solvent such as alcohol is used as a granulating fluid.

Drying

The most common conventional method for drying a granulation continues to be the circulating hot air oven, which is heated by either steam or electricity. The important factor is to consider as part of scale-up of an oven drying operation are airflow, air temperature, and the depth of the granulation on the trays. If the granulation bed is too deep or too dense, the drying process will be inefficient, and if soluble dyes are involved, the migration of the dye to the surface of the granules. Drying times at specified temperatures and airflow rates must be established for each product, and for each oven load. Fluidized bed dryers are an attractive alternative to the circulating hot air ovens. The important factors considered as part of the scaled - up fluidized bed dryer are optimum loads, rate of airflow, inlet air temperature and humidity.

Reduction in Particle Size

The first step in this process is to determine the particle size distribution of granulation using a series of "stacked" sieves of decreasing mesh openings. The particle size reduction of the dried granulation of production size batches can be carried out by passing all the material through an oscillating granulator, a hammer mill, a mechanical sieving device, or in some cases, a screening device. As part of the scale-up of a milling or sieving operation, the lubricants and glidants, in the laboratory are usually added directly to the final blend. This is done because some of these additives, especially magnesium stearate, tend to agglomerate when added in large quantities to the granulation in a blender.

Blending

The type of blending equipment often differs from that using in laboratory scale. In any blending operation, both segregation and mixing simultaneously occur simultaneously as a function of particle size, shape, hardness, and density, and of the dynamics of the mixing action. Particle abrasion is more likely to occur when high-shear mixers with spiral screws or blades are used. When a low dose active ingredient is to be blended, it may be sandwiched between two portions of directly compressible excipients to avoid loss to the surface of the blender.

Slugging

In the dry granulation method, the powder-mix is compacted in the form of slugs. Slugs range in diameter from 1 inch, for the more easily slugged material, to ¾ inch in diameter for materials that are more difficult to compress and require more pressure per unit area to yield satisfactory compacts. This is done on a tablet press designed for slugging, which operates at pressures of about 15 tons, compared with a normal tablet press, which operates at a pressure of 4 tons or less. If an excessive amount of fine powder is generated during the milling operation, the material must be screened and fines recycled through the slugging operation.

Dry Compaction

Granulation by dry compaction can also be achieved by passing powders between two rollers that compact the material at pressure of up to 10 tons per linear inch. Materials of very low density require roller compaction to achieve a bulk density sufficient to allow encapsulation or compression. One of the best examples of this process is the densification of aluminum hydroxide. Pilot plant personnel should determine whether the final drug blend or the active ingredient could be more efficiently processed in this manner than by conventional processing to produce a granulation with the required tableting or encapsulation properties.

Compression

The ultimate test of a tablet formulation and granulation process is whether the granulation can be compressed on a high-speed tablet press. When evaluating the compression characteristics of a particular formulation, prolonged trial runs at press speeds equal to that to be used in normal production should be tried, only then are potential problems such as sticking to the punch surface, tablet hardness, capping, and weight variation detected. Highspeed tablet compression depends on the ability of the press to interact with granulation. The following parameters are optimized during pilot plant techniques of granulation feed rate; delivery system should not change the particle size distribution. The system should not cause segregation of course and fine particles, nor should it induce static charges. The die feed system must be able to fill the die cavities adequately in a short period that the die is passing under the feed frame. The smaller the tablet, the more difficult it is to get a uniform fills a high press speed. For high-speed machines, induced die feed systems are necessary.

These are available with various feed paddles and with variable speed capabilities; so that the optimum feed for every granulation is obtained. Compression of the granules usually occurs as a single event as the heads of the punches pass over the lower and under the upper pressure rollers. This causes the punches to penetrate the die to a pre-set depth, compact the granules to the thickness of the gap set between the upper punch and lower punch. During compression, the granules are compacted to form tablet. Bonding among the compressible materials, due to cohesive force, must be formed, which results in the formation of tablet. Along with the cohesive force, adhesive force between granule and punch surface may be developed. If the strength of the adhesive force is more than the cohesive force, sticking may occur. This happens particularly when an embossed punch is used. To overcome sticking, the granules must be lubricated internally. If the granules contain a higher amount of lubricant or are lubricated for a longer time (over blending) soft tablets are produced, the wettability of the powder is decreased, and the dissolution time of the tablet can increase.

Binding to die walls can also be overcome by designing the die to be 0.001 to 0.005 inch wider at the upper portion than at the centre to relieve pressure during ejection. The machines used are high speed rotary machine, multirotary machine, double rotary machine, upper punch and lower punch machine and single rotary machine.

Whatever may be the tablet compression machine, during compression it performs at least three activities;

- Filling of empty die cavity with blended granules
- Compression of granules into tablets, and
- The ejection of tablets formed from the die cavity.

Sometimes a greater number of trials should be run during scale-up process. Safe operation, the number of tablets produced in unit time (productivity) and quality of tablets produced are to be optimized and for these, following points must be considered during pilot plant scale up technology.

- Characteristics of raw materials and/or formulation constraints are not suitable for high speed compression. The compression speed is reduced or tablet press of lower speed is used; so that the time required for filling the die cavities and compression can be optimized to achieve sufficient compaction process.

- Selection of suitable tablet press is necessary for a particular formulation; because the tablets compressed must be free from any defects and must be strong enough to withstand the shock of shipment and transportation. The tablet press with improved feed control mechanism, precompression and compression forces can solve many problems related to compression.

- Some granules have a tendency to cap when subjected to a single-step compression. This problem can be effectively solved using a tablet press with a series of pressure rollers that increase the compaction pressure stepwise; so that the air entrapped within the granules is escaped gradually and at the end capping problem would be over.

- High speed tablet press requires delivering the granules into the die cavities at a consistent and required rate. The rate of die filling should not change until the compression is completed.

- During die filling, theparticle size distribution in the granules must not change for any reason. This is because the uniformity of the drug content may change if the original particle size distribution is changed.

- The development of a static charge on the particles is another common problem that occurs frequently when segregation or separation of particles occurs particularly in case of small tablets. The die cavity must be filled uniformly with granules. Hence, particle size distribution and particle flow property should be good enough and the mean particle size should be small to ensure uniform filling of the dies.

- If the die cavities are overfilled, the excess granules need to be removed by the feed frame to the centre of the die table. Otherwise, the granules shall be thrown from the table by the centrifugal force of the rotating die table. Hence, the scraper blade and die table should be adjusted to make necessary clearance for the removal of excess granules.

- Nowadays, high speed tablet press with induced die feed systems have been developed. These are also available with variable speed and different feed paddle. Such a system can feed every type of granule at an optimum level.

- During compression the materials present in granules are bonded, compacted, and finally the tablets are produced. A strong cohesive force within the material exists at the tablet interfaces. Along with the cohesive force adhesive force is also developed between tablet interfaces and die surface and punch. If the adhesive force is more than the cohesive force sticking of tablets will occur. Therefore, to avoid sticking adequate lubrication of the granules is required. Generally, metal stearates such as magnesium stearate are most widely used. The amount of lubricant to be used should be optimized; because poor lubrication may cause sticking, while excess lubrication and blending for a longer period may result in other problems such as softening of the tablets, poor wettability, and increase in dissolution time.

- The last step of the compression process is the ejection of tablets from the die cavity. In this step, the upper punch goes up leaving the tablet and the lower lunch push up the tablet to the die table. The take-off bar removes the tablet from the rotating die table. The tablet is collected through the chute fixed at a particular position adjacent to die table. The take-off bar and chute should be adjusted in such a way that the tablet once removed by bar falls into the chute; because the die table rotates.

The design and condition of the punch can cause sticking also. For example, embossing with punch can cause sticking. The punches and dies must be cleaned and kept carefully. The punch tips must remain sharp and smooth.

Tablet Coating

Since long the sugar coating has been used to coat the tablets using pan coating method. With development of technology, polymers and equipment film coating has replaced the sugar coating in many cases. The conventional pan has been replaced by perforated pan and fluidized bed coating column. Earlier polymers soluble in volatile inflammable organic solvents had been used. Due to safety problem and health hazards the earlier polymers have been replaced by newly developed water-soluble polymers.

As such, coating is a specialized activity. In manual system, the finish of coat depends on the expertise of the operator. Even after development in laboratory scale, it requires to be standardized or optimized during commercial production. Quality and finish of the coat depends mainly on;

- Design of the core tablet – the edge of tablet should not be sharp and surfaces should not be flat. Tablets with curve, shallow or deep, with bland edge are suitable for coating.

- Strength of the tablet – the tablets should be hard enough to withstand the shock produced during tumbling or rolling in the coating pan or in fluidized coating column.

- Embossing or engraving on tablet surface – the edge of the engraving should not be sharp and deep. The engravings should be shallow and the edges are bland or cuts should be angled.

- Nature of the core material –hydrophobicity of the core material is not suitable for aqueous coating solution. If the core components are originally hydrophobic, either the composition of the core or of the coating solution is be modified.

- Lot size of the core tablets – coating quality on small lot size may not match with that of higher lot size. This is due to increase in number of tablets (lot size), environmental conditions (temperature and humidity), rate of application of coating solution and drying cycles. Hence, further trials on larger batch size should be made to standardize the method.

- Friability or brittleness of the tablets –brittle tablets are not suitable for coating. The tablet should not break into fractions or should not produce powder due abrasion during rolling. If so happens, the surface of coated tablets would be rough and twin tablets would be produced.

- Design of the coating pan – chipping and loss due to abrasion can be appreciably reduced if the conventional coating pan is provided with baffles. Due to these baffles the tablets can roll uniformly; but not slide. The tablet load would be redistributed and spread more uniformly over the entire tablet weight.

- Design of the coating column – by changing the length and diameter of the column of the fluidized bed coating column, the loss due to abrasion can be reduced appreciably. In this method the force of fluidizing air needs to be controlled.

- Design and type of the nozzle – for coating of tablets either the air-atomizing nozzle or airless atomizing nozzle can be used. If an airless sprayer is used the size and shape of nozzle is important. In case of air-atomized sprayer, the atomizing air pressure and the solution flow rate are important.

In traditional coating system the pan is attached to an exhaust duct to collect dust. As such the process creates a lot of dust and a single operator can handle four to five pans. The pan may be open or closed one. During coating lot of noise is also produced. Nowadays automated computerized method with microprocessor has been developed to coat the tablets.

Filling of Hard Gelatin Capsules

This has already been mentioned in the general method of manufacture of capsules that each size of capsule has a definite fill volume. Thus, the powder to be filled in a particular capsule shell must have definite bulk density, so that the required amount can be filled in a particular capsule. Depending on the amount of drug (unit dose of the drug) to be filled, its bulk density and capsule size, the drug can either be filled as such or first processed and then filled in the empty capsules. Hence, the drug is mixed with other excipients as is done to make tablets, mixed, and the dry powder blend is filled in capsules. The powder mix is wet granulated using a suitable binder; wet granules are dried, sieved and mixed with suitable lubricant. The lubricated granules are then filled in capsules. Thus, the manufacture of both capsule and tablet has got similarity.

For high speed machinery the powder/granules must have the required bulk density, particle size distribution, good flow property, and adequate compressibility for filling the capsules. The required number of particles should be compacted within the capsule shell to ensure filling of the correct amount of drug in each capsule. The excess lubricant present in the granules/powder or over-lubrication may increase the disintegration time and reduce the rate of dissolution. Environmental conditions are also to be considered; because definite temperature and humidity are required for the stability of the capsule shell. Hence, the factors to be considered during encapsulation of hard gelatin capsules are;

- Bulk density of powder/granules
- Particle size distribution
- Moisture content of the powder
- Powder/particle flow properties
- Compressibility of powder/granules
- Lubrication of powder/granules, and
- Environmental conditions.

Thus, during the pilot plant scale up process sufficient trials must standardize the filling operation. Usually the temperature should be maintained within 15 to 25°C and relative humidity within 35%-55%. If the humidity is high the capsule will swell because it absorbs moisture; if the humidity is too low the capsule shell will release moisture and will become brittle.

Liquid Dosage Forms

Liquid dosage forms commonly refer to nonsterile liquid dosage forms that include solution, suspension, and emulsion. These are either administered orally or topically. The scale - up of each product should be considered separately. Sterile liquids constitute a special class of preparations and require a separate consideration.

Solution

Solutions belong to a simple class of formulations and its scale up technique is also simple. Container/tank of suitable size and a mixer of required power and capacity are required for scaling up of the manufacturing process. For faster solubilization of solutes, heating facility should be attached to the tank. Generally, steam jacketed tank is used to prepare the solution or mixing. The solution requires filtration and transportation. Hence, suitable filtration equipment such as sparkler filter (filter press) is used for filtration of bulk solution. For transportation of solution of suitable capacity is used. All the equipment's used must be made of nonreactive material and should be cleaned easily. Stainless steel of type 308 or 316 is usually used to construct the equipment's, pipe, etc. Commonly stainless steel of type 316 is used because it is less reactive in nature. In fact, stainless steel is not nonreactive, it is less reactive. Some acidic preparations react with stainless steel. In such case, the steel should be treated with dilute acetic acid or nitric acid to neutralize the surface alkali of the steel. This process should be repeated on regular interval.

This process is called *passivation*. This must be out immediately after the preparation of an alkaline solution. Otherwise, the inner metallic surface should be coated with glass or polytetrafluoroethylene (Teflon). The metal surface becomes completely nonreactive, but it has disadvantage like cracking, flaking, peeling, or breaking.

Suspension

In case of disperse system the scale up ratio generally varies, from 10 to 100 from laboratory scale to pilot scale and from pilot scale to commercial scale varies from 10 to 200.

For scaling of a suspension formulation is not as simple as a solution. The disperse systems except microemulsion are in general thermodynamically unstable, multiphase system. Interfacial phenomena play an important role in making the system stable. The heterogeneity of the interphases is important, particularly due to the presence of solid particles. The solid state is the most complex state of matter and it confuses the process translation. The viscosity and flow properties of a suspension may vary considerably during the process translation. For example, in, the laboratory, the suspending agent is sprinkled over water and then stirred to prepare its dispersion. This is impossible in a large-scale production. In large-scale production, the dispersion medium is prepared using a vibrating feed system. Usually the steps in making a suspension are;

➢ Mixing

➢ Particle size reduction

➢ Material transfer

➢ Heat transfer

Mixing

This is the basic operation of suspension manufacturing. Sometimes the term agitation is used in place of mixing, which is wrong. When initially separated two or more phases are randomly distributed into and through one another, the process is called mixing or blending; while agitation causes movement or rotation of a material into a container. Agitation may not mix two or more separate components of a system into a uniform mixture.

Depending on the blades attached, a mixer may be propeller, turbine, paddle, helical ribbons, Z-blades, or screws. Moreover, the number of impellers, number of blades per impeller, the pitch of the impeller blades, and the location of impeller can be changed to change the performance of the mixer. However, for effective mixing, a dispensator or rotor/stator mixer can be used in place of an impeller.

The manufacture of a suspension or emulsion mixing process may be a problematic operation due to the higher consistency and non-Newtonian flow characteristics of the system. Both laminar and turbulent flows occur simultaneously in different regions of the system. This does not happen in the case of low viscous liquids mixing. In high viscosity materials (η > -10^4cps) mixing is relatively slow and is found not so effective. Generally, in such mixing, laminar flow occurs in place of turbulent flow; as a result, the inertial forces passed on during mixing disperse fast. Efficient mixing is required to provide convective flow. This is possible by high velocity gradient in the mixing zone. Hence, for highly viscous materials, mixing with specialized impellers, turbines, paddles, anchors, helical ribbons, and screws is suitable. Thus, during scaling up the proper mixing equipment should be selected.

Particle Size Reduction

Generally, the suspension is the dispersion of solid particles in a liquid vehicle. On storage, the dispersed particles may agglomerate, which is not desired. To produce homogeneous suspension and for prolonged dispersion, the dispersed solid particles should be very fine; hence, the size of the solid material should be reduced. With a decrease in the particle size, the number of particles increases. Thus, the interparticle and interfacial interactions increases. Particle size can be reduced by any of the four methods– (1) compression, (2) attrition, (3) impact, and (4) cutting or shear using particular equipment. Each method works with a particular mechanism.

- Crushing roll works involving a compression mechanism
- Grinder such as hammers mill, ball mills operate primarily with both the mechanism impact and attrition; however, compression is partially involved
- Ultrafine grinders such as fluid energy mills work primarily with attrition.

The grinding or crushing efficiency, $E_{C,}$ is the ratio of surface energy created by crushing or grinding to the energy absorbed by the solid and can be expressed as;

$$E_C = \frac{\sigma_s}{w_n}\left(A_{wp} - A_{wf}\right)$$

where, σ_s is the specific surface the area, A_{wp} is area per unit mass of product particles, A_{wf} is the area per unit mass of feed particles and W_n is the energy absorbed by the solid per unit mass. If W is the energy supplied to the mill; then $W - W_n$ is the energy spent for friction during crushing.

However, milling of the disperse phase using a ball mill, roller mill, and fluid energy mill has been found successful.

Usually the viscosity of dispersion is greater than that of its equation dispersion medium and according to the Einstein's equation, this can be expressed as;

$$\eta = \eta_o(1 + 2.5\varphi)$$

where, η is the viscosity of the dispersion, η_o is the viscosity of the dispersion medium, and φ is the volume fraction of the particular phase. The viscoelasticity of pharmaceutical suspensions indicates that these are non-Newtonian fluids – plastic, pseudoplastic. Their rheological behavior depends on φ; more precisely due to deformation of disperse phase and/or interparticulate interaction.

Material Transfer

The effect of the rate of mass transfer on the changes in the properties of a product is often neglected or ignored during scaling up of a manufacturing process. But it is an important activity and should be considered. During the manufacture of suspension, the material is transferred from the mixing or holding tank to processing equipment or to the filling line either by pumping or by gravity feed. This operation in most of the cases is found potential problematic. During pouring or transfer through the pump change in particle size distribution, physical or chemical instability may occur. Such changes occur due to the changes in the rate of transfer or in shear rate or shear stress. Hence at the time in scaling up of material transfer operation such changes in suspension and its time dependency should be considered and optimized because the product should be uniform at every point in time. Hence, during commercial production mass transfer time and its impact on product's quality must be scaled up.

Heat Transfer

In most of the cases heat is required to dissolve the materials while making a liquid formulation. Heat transfer is thus, an important parameter in the production of pharmaceutical formulations. The rate of heat transfer depends on the ratio of volume to surface area. This ratio is relatively small in laboratory scale manufacturing; as a result, heat is transferred quickly and the product is cooled relatively faster. This is not true in the case of large-scale production. In large scale and even on pilot scale this ratio is large; hence, the time taken for cooling of the product is longer. As a result, the product is exposed to higher temperatures for a long time and may undergo thermal degradation. Thus, the mixing tank or solution preparation tank should be jacketed one through which cold water can be circulated to cool the product rapidly and if required, steam can be circulated in the jacket to increase the temperature of the contents in the tank.

The rate of heat transfer can be calculated as follows;

Say, $\dfrac{dQ}{dt}$ is the rate of heat transfer, A is the heated surface of jacketed tank, T_o is the temperature maintained within the jacket and T is the temperature of content in the tank, then $\dfrac{dQ}{dt}$ is proportional to heated surface area, A and the temperature gradient, $(T_o - T)$ at time t.

$$\frac{dQ}{dt} = C_p \left(\frac{dQ}{dt} \right)$$

where, C_p is the heat capacity of the jacketed vessel and its contents. If k is the heat transfer coefficient, then

$$\frac{dQ}{dt} = kA\,(T_o - T)$$

If $\dfrac{kA}{C_p} = a$ and T_i is the initial temperature of the tank, the above equation can be expressed as;

$$T_o - T = (T_o - T_i)\,e^{-at}$$

Thus, by using the above equation, the time, t required can be calculated to bring down the temperature, T_f, of the content in the tank.

Thus, on the basis of scale up performance the type of mixing tank, mixer, pump, mill, and the horse power of the motor should be carefully selected to prepare a commercial batch of pharmaceutical suspensions. The size of the equipment's should be selected as per commercial batch size. The type of mixer should be as per the maximum viscosity of the product.

According to Kline et al. The approaches to scale up of disperse systems can be classified into three categories;

- Modular scale up or limited scale-up
- Known scale up or limited scale-up, and
- Fundamental approach when scale-up ratio is high

Modular scale up refers to the scaling up of individual components or unit operations of a manufacturing process. The relationships among these individual operations result in a major scale-up problem. That is, if the process is used in large scale, the same quality of the product is not achieved. However, with the knowledge of the physicochemical properties of dispersion components, it may be possible to predict the scalability of some unit operations.

When laboratory or pilot scale experiments are less, known scale-up correlation can be applied using mathematical modeling of the manufacturing process and at various scale-up ratios experimental validation of the model can be done.

In fact, for scale-up of the dispersed system, no virtual guideline or model is available. The trial-and-error method has been commonly practiced and for success in the scalability of the process, practical experience plays a major role.

Emulsions

Emulsions belong to the disperse system, in which the globules of a liquid remain dispersed in a dispersion medium. The liquid dispersed phase may be a pure liquid or a solution. The emulsions may be O/W or W/O. In oil-in-water emulsions, oils or waxes are dispersed as globules in water while in water-in-oil emulsions aqueous solution remains dispersed globules in an oil medium. Emulsions may be in liquid form or in semisolid form, such as cream, ointment, etc. The manufacture of a liquid emulsion is a specialized process; hence, during scaling-up of the process selection of appropriate equipment and optimization of process parameters is very crucial. Extensive process development work and validation would be required.

The process parameters include the type of mixer, mixing speed, temperature, type of homogenizer, cooling process, screens, pumps used for material transfer, and filling equipment, etc. The physical properties such as appearance and viscosity, and physical stability of emulsion depend on how and to what extent the globule size of the dispersed phase has been reduced. The use of high-shear mixer generally causes air entrapment and therefore, physical appearance and chemical the stability of the product may greatly be affected. The use of a mixing container operated under the controlled vacuum does not require aeration; as a result, above mentioned problems can be avoided. The presence of particulate matter in an emulsion affects the quality of the emulsion. Hence, the filtration of an emulsion is necessary to remove the foreign particles. It is better to filter the oil phase and water phase separately before emulsification, and sufficient care should be taken to ensure that no particulate matter is introduced into the product during preparation. It is wise to avoid filtering the final product.

To avoid, the problems of scaling-up and to answer the questions that may arise after the process has been operating smoothly for months or even years the pilot plant must be well designed. Thus, during pilot plant manufacture the same type of equipment's of smaller size must be used. In many cases scaling-up experiments are difficult to conduct with small volume and if done, it becomes very expensive. The equipment's should be capable of making a product of 20 to 100 lt. Even after scaling-up and successful production of few batches, it may be necessity to standardize the method of production. For example, any change in material supplier or change in personnel or change in the performance of a particular machine, there may be some deviation in the final quality of the product; hence, it becomes necessary to find out the reason responsible for such change in quality. It is expected that there will be no change in the production procedure.

Semisolid Dosage Form

Semisolids include paste, ointment, and cream. These also belong to dispersed system like suspension and emulsion; but their viscosity is very high. Thus, many Scale-up parameters related to these preparations are similar to suspension and emulsion. Because of high viscosity some parameters of these products are critical in nature. For example, mixer used to prepare suspension or emulsion would not be useful for ointments and creams. The turbulence created by the mixer during mixing of suspension shall entrap air which would be very difficult to remove. The mixing operation in making ointment should be such that the mass is removed effectively and continuously from side wall of the container to central zone and from bottom to top. The speed of impeller should be controlled to ensure uniform mixing without or minimum entrapment of air.

Processing steps such as during emulsification the mixing of aqueous and oil phases, homogenization, addition of active ingredient, and product transfer are usually performed at predetermined temperature. The container or tank should be jacketed to circulate steam to heat the mass in the tank. After thorough mixing, the preparation needs to be cooled by circulating cold water. The temperature at which these operations are carried out is critical to the quality of the final product. For emulsification the temperatures of both aqueous and oil phases are raised to a temperature higher than the solidification point of the oil phase. This temperature is important for emulsification and the quality of the final product. The rate of heat transfer should be adjusted in such a way that the product is cooled slowly without affecting the texture of the product. Improper emulsification may result due to improper temperature control and viscosity differences in products. The particle size of poorly soluble drug can also be greatly affected by improper temperature control. If the drug is added when the temperature of the emulsion is too high, the solubility of the drug increases, a saturated solution of the drug is formed with the formation of a metastable product. After that, if the product is cooled recrystallization or crystal growth may take place, the crystals may be of different size or polymorphic form and particle size distribution can change. Finally, a gritty, less elegant, less stable product is formed. The product may be less therapeutically active.

Some gel and cream formulations are sensitive to shear. Such products should be manufactured carefully. These products should not be exposed to high shear during transportation from the manufacturing container to the holding container or filling lines. The viscosity of the product may change during such transportation or filtration, which is undesirable. Another problem occurs during the homogenization of an emulsion. For this purpose, various types of high shear mixer such as homogenizer, and colloid mill, are used. Among these, the colloid mill is the most commonly used. A colloid mill contains a fixed stator plate and a rotor platerotates at high speed. The gap between the stator and rotor is set to reduce the particle size of the product by the physical action and centrifugal force exerted by the rotor. The gap is adjusted within 0.005 to 0.01 inch. Recently supersonic homogenizer is used to reduce the particle size or aggregates and to micronize.

Highly viscous semisolid product is usually transferred using a positive displacement pump. This pump works without applying excessive shear or entrapment of air. The selection of pump should be based on the viscosity of the product, the rate of transfer, compatibility of the product with the pump surface, and the pumping pressure.

Thus, for scaling-up the manufacturing method of a semisolid product following processing parameters should be considered.

- Temperature at which emulsification to be done
- Viscosity of the product
- Temperatures of the aqueous and oil phases
- Rate of cooling the emulsion
- Homogenizer
- Selection of pump for transfer of product for filling

Relevant Documentation

During R&D processes before technology transfer of drug substances, information indicated below should be collected, and based on the information, technology transfer documentation should be prepared.

Information related to Raw Materials, Intermediates and Drug Substances

➢ Impurity profile and information on residual solvents (structure of impurities and route of synthesis)

➢ Information on descriptions of crystals of drug substances (crystallization, salt and properties of powders)

➢ Information on stability and description (raw materials, drug substances (including packaged drug substances), intermediates, solutions, crystal slurry, and humid crystals)

➢ Information on safety of drug substances, intermediates, and raw materials (Material Safety Data Sheet (MSDS))

➢ Information on animal origins of raw materials, etc.

➢ Information on packaging materials and storage methods (quality of packaging materials, storage temperature, and humidity)

➢ Information on reference standards and seed crystals (method of dispensing, specifications and test methods, and storage methods)

Information related to Manufacturing Methods

➢ Information on manufacturing methods (synthetic routes and purification methods)

➢ Information on operating conditions (control parameters and acceptable range)

➢ Information on important processes and parameters (identification of processes and

- Parameters, which will affect quality)
- Information on in-process control
- Information on reprocess and rework (places and methods)
- Basic data concerning manufacture (properties, heat release rate, reaction rate, and solubility, etc.)
- Data concerning environment and safety (environmental load and process safety)

Information related to Facilities and Equipments

- Information on equipment cleaning (cleaning methods, cleaning solvents, and sampling methods)
- Information on facilities (selection of materials, capacity, and equipment types, and necessity of special equipments)

Information related to Test Methods and Specifications

- Information on specifications and test methods of drug substances, intermediates, and raw materials (physical and chemical, microbiological, endotoxin and physicochemical properties, etc.)
- Validation for test methods of drug substances and intermediates

Items to be verified for Review of Scale-up

Sometimes manufacturing processes of drug substances use unstable chemical substances, and the unsteady processes accompany with chemical changes. Hence, handling period for each operational unit, stability of subject compounds during operation, and conditions of scale-up should be carefully established.

Since, the parameters related to equipments, operations can influence the quality of the product, they should be considered carefully.

Items to be verified for Reviewing Scale-Up of Reaction Processes

- Reproducibility of temperature pattern and its effects (effects of delay in temperature up and down on quality)
- Effects of churning in heterogeneous and semi-batch reactions (formations of concentration distribution and diffusion-controlled zone)
- Prediction and effect of operation period of consecutive reactions or exothermic reactions in semi-batch reactors (extension of operation

period due to insufficient capacity of facilities and its effects on quality)

- Balance between heat release rate and heat dissipation capacity (temperature pattern of exothermic reaction and its effects)
- Effects of facilities (validity of required capacity of utility, and effects from temperature distribution, dead volume, and overheating of laminar film)
- Confirmation of fluctuations due to scale-up (phenomenon, which did not appear at flask levels)

Items to be confirmed in Reviewing Scale-up of Crystallization Processes

- Effects of churning (effects on particle size and polymorphism, and selection of scale-up factors)
- Reproducibility of temperature pattern (reproducibility of established temperature pattern and effects on quality)
- Effects on facilities (temperature pattern, changes in flow pattern, effects of local concentration distribution and temperature distribution, super-cooling of laminar film, and scaling)
- Prediction of time for solid-liquid separation and its effects (stability of slurry waiting for filtration)
- Confirmation of ease of operation (problems at actual equipment levels such as slurryemission, transfer, and churn load)

Clarification of Quality Variability Factors

To elucidate quality variability factors, the following items should be reviewed during quality design through scale-up review.

Processes Affecting Quality

To identify processes, which may affect the quality of final drug substances, such as processes to generate final substances, structures with pharmacological activities, and impurities that cannot be eliminated in purification.

Establishment of Critical Parameters Affecting Quality

To search for parameters among those controlling the above processes, which may affect the quality of final drug substances, such as generation and elimination of impurities, and physicochemical properties of final drug substances, and establish a range of control.

Establishment of other Parameters

Parameters not affecting the quality of final drug substances are not subject to validation; however, they are subject to change control and change histories should be recorded.

Development Report on Synthetic Drug Substances

Information concerning the drug substances and intermediates to be described is as follows:

- Development history, including different synthetic methods used to manufacture investigational drugs
- Finally determined chemical synthetic route
- Change history of processes
- Quality profile of manufactured batches
- Specifications and test methods of intermediates and final drug substances
- Rationale for establishment of critical processes
- Critical parameters and control range
- References to existing reports and literature, etc.
- Master batch records of the manufacturing of investigational drugs or samples (format of manufacturing records)
- Manufacturing records of investigational drugs or samples (batches for establishment of specifications for application, validation batches, etc.)
- Plan and report of process validations
- Items of in-process control: IPC (test methods and specifications)
- Investigation report on causes of abnormalities (if occurred)

Information on Cleaning Procedures
- Master batch recording of cleaning
- Record of cleaning
- Plan and report of cleaning validation
- Test methods and specifications
- Validation report on analytical methods used for cleaning validation

Information on Analytical Methods

- Development report on analytical methods or those corresponding to the development report
- Test methods and specifications (raw materials, intermediates, final drug substances, and container/closure)

➤ Validation report on release test methods

➤ Stability test (validation report on analytical method, plan/report of stability test, container form, reference standard, and relevant reports)

➤ Investigation report on causes of OOS (out of specification)

Information on Methods of Storage/Transportation

➤ Container/closure system

➤ Date of retest/expiry date

➤ Conditions of transportation

➤ Information on sensitivity to temperature, humidity, light, and oxygen

➤ Instructions for temperature monitoring of drug substances which need cold storage

Information on Facilities

➤ Structural materials

➤ Category and type of main facility

➤ Critical facilities for final processing that may affect physicochemical properties (particlesize and surface conditions, etc.)

SUPAC Guidelines

In 1991, the American Association of Pharmaceutical Scientists, the Food & Drug Association (FDA), and the United States Pharmacopeia (USP) jointly organized two workshops to investigate the aspects of Scale-Up and Post-approval Change (SUPAC) in immediate-release of oral solid dosage forms and in 1992for oral extended-release dosage forms. The decisions taken in both workshops were published in 1993 and have been used as guidance to the industry and regulatory bodies.

Primarily four aspects discussed and defined in those proceedings are:

1. The impact of changes in formulation and composition on the finished quality parameters of these products.

2. The impact of changes in the process variables on the finished quality parameters of these products.

3. The impact of changes in process scales on the finished quality parameters of these products.

4. The impact of changes in process site on the finished quality parameters of these products.

To assess the significance and to record and report the post-approval changes, the FDA has issued the guidance for SUPAC changes designated as the following:

- Supac-IR (immediate release solid oral dosage form)
- Supac-MR (modified release solid oral dosage form)
- Supac-SS (non-sterile semi solid dosage form such as cream, ointment, gel and lotion)

These guidelines provide recommendations to the pharmaceutical industry producing new drug applications (NDAs), abbreviated new drug applications (ANDAs), and abbreviated antibiotic drug applications (AADAs). Sometimes, the manufacturers wish to change:

(a) The components or composition,

(b) The site of manufacture,

(c) The scale-up/scale-down of manufacture, and/or

(d) The manufacturing (process and equipment) of a modified releaseor immediate release solid oral dosage form during the postapproval period.

The guideline defines:

1. Levels of change such level 1 refers to minor changes, level 2 refers to moderate changes, and level 3 refers to major changes.

2. Recommended chemistry, manufacturing, and controls (CMC) tests for each level of change,

3. Recommended in vitro dissolution tests and/or in vivo bioequivalence tests for each level of change; and

4. Documents supporting the change.

Level 1: when the changes are unlikely to have any measurable/detectable effect on quality and performance of the formulation. For example, (1) change in colon, flavor. (2) Changes in the amounts of excipients (expressed as the percentage (%w/w) of total of unit product weight) do not exceed5%.

Level 2: when the changes may have a significant effect on the quality and performance of the formulation. For example, (1) change in the technical grade of the excipient, such as use of Avicel PH102 in place of Avicel 200 or vice versa. (2) When changes in the number of excipients (expressed as the percentage (%w/w) of total of unit product weight), exceeds 5%, but not more than 10%.

The guidelines specify that an application should be given to the Centre for Drug Evaluation and Research (CDER). It should provide the information to ensure the product quality and performance characteristics of a modified release or immediate release solid oral dose formulation for specified postapproval changes.

Level 3: when the changes are likely to have a significant effect on the quality and performance of the formulation. For example, (1) qualitative or quantitative change of excipient mixed with a narrow therapeutic drug beyond the range of Level 1. (2) When the dissolution criteria of a drug do not meet the specification of Level 2.

Extended release solid oral dosage forms:

Level	Classification	Therapeutic range	Test Documentation	Filling Documentation
I	Complete or partial deletion of colour/flavour Change in inks, imprints Up to Supac-IR level 1 excipient ranges No other changes	All drugs	• Stability • Application/compendial requirements • No biostudies	Annual report
II	Change in technical grade and/or specifications Higher than Supac-IR Level1 but less than Level2 excipient ranges No other changes	All drugs	• Notification and updated batch records • Application/compendial requirements plus multi-point dissolution profile in three other media (water, 0.1N HCl, USP buffer media pH 4.5, 6.8) until $\geq 80\%$ of drug is released or an asymptote is reached. • Apply some statistical test (F2 test) for comparing dissolution profiles • No Biostudy	Prior approval supplement

Contd....

Level	Classification	Therapeutic range	Test Documentation	Filling Documentation
III	Higher than Supac-IR Level2 excipient ranges	All drugs	• Updated batch record • Stability • Application/compendial (profile) requirements • Biostudy or IVIVC	Prior approval supplement

Introduction to Platform Technology

In the 1980s, the development of monoclonal antibodies (mAb) was started. Since then, it has become vastly successful as an effective and safe therapeutic for various diseases. The idea of establishing and applying Platform Technology for a process is still new and still there is no definition in ICH Q18 (R2). Now the platform technology has become one of the fastest growing areas in the pharmaceutical industry.

For each molecule development of antibodies was a tailored approach and this happened from discovery to development. Attempts have been made to establish a definite process and to assure quality by compliance. The knowledge about the quality of the product and technology started to increase the result development of new technology to achieve the safety and efficacy of biologics with the concept of Quality-by-Design (QbD). By adding data for every new molecule developed using this method, the platform can be improved continuously.

However, according to the principle of QbD, Platform Technology can be defined as; a systematic approach to influence the prior knowledge for standardized processes that has been revealed that the multidimensional combination and interaction of input variables and process parameters can be applied for a class of molecules with comparable characteristics to achieve assurance of quality.

According to this definition, it is necessary to establish the platform technology for the identification of a class of molecules possessing similar or comparable characteristics with respect to their physico-chemical properties and stability such as monoclonal antibodies. Any molecule having characteristics can be considered to belong to the same class.

Prerequisites to establish a Technology Platform for Early Development

To establish and justify the applicability of a Platform Technology, it is necessary to identify the molecules which can be considered next-in-class molecules.

A combined effort of multiple functions in late-stage discovery, involving candidate generation, screening and selection of the lead candidates is necessary to determine whether the selected lead molecule belongs to next-in-class molecule or not. This would be the systematic approach for this.

Target Molecule Profile

The characteristics of a drug molecule should be similar to those of a product. To develop a target drug product, thetarget product profile needs to be considered. Hence, the Target Product Profile must be defined first, and as per this definition a Target Molecule Profile (TMP) should be searched and the important or necessary characteristics of a molecule should be specified to ensure stability, safety and therapeutic efficacy. The figure 5 shows the screening process of potential target molecule profile, assessment of their drug like properties and to justify whether the platform process can be used on these molecules.

Screening and Selection

Such a screening process involves several functions and is carried out in different stages. The method can be used for a wide number of materials for screening at each stage. At the first stage assay is usually done to evaluate the primary amino acid sequence and to screen the methods to identify the sequence and structural configurations that might cause chemical or physical degradation, such as degradation, aggregation, colloidal instability or unwanted biological effects like immunogenicity.

The accuracy of this screening method cannot eliminate the candidates solely; subsequent testing is necessary to assign a development risk to each molecule. Based on prior knowledge, the molecules can be ranked according to predefined benchmark criteria. The molecules showing the overall best drug-like properties are selected as lead and backup candidates.

Finally, the result of the screening justifies whether the selected lead and backup candidate is a next-in-class molecule, and can be subjected to the platform process.

If none of the screened molecules match with the requirements of a next-in-class molecule, more extensive and customized formulation and process development efforts are made.

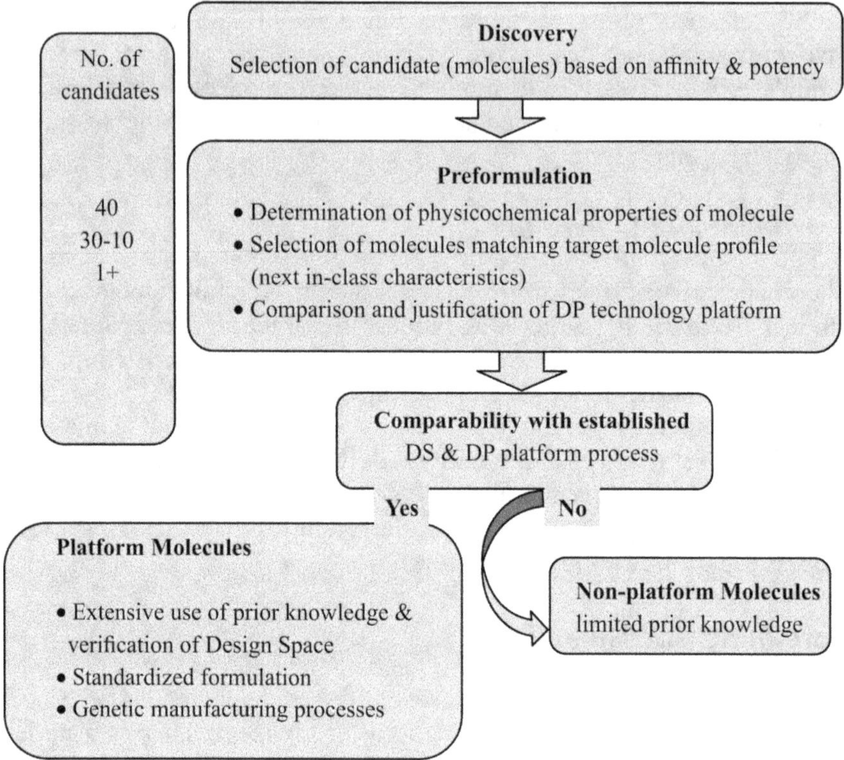

Fig. 1.6 Flow-chart of a standardized screening process to assess and identify next-in class molecules as defined by the Target Molecular Profile (TMP)

Establishing a Platform Process in Early Drug Product Development

The previously described activities are performed either after or during the late-stage discovery. This section provides an indication of the activities to be carried out in the development area.

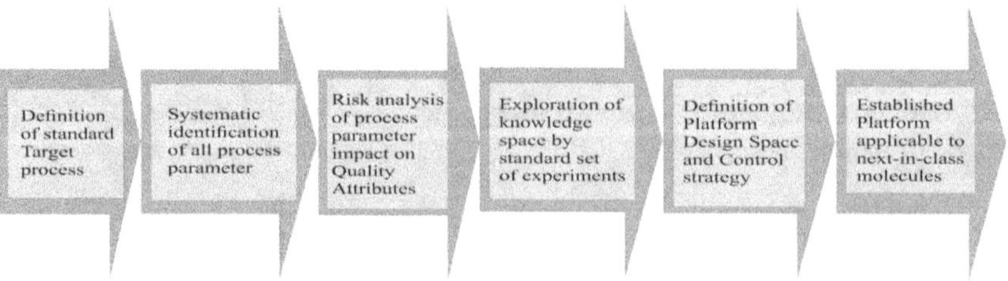

Fig. 1.7 Flow-chart of the various stages of implementing a technology platform

As shown in Figure 1.7, the method used to set up the platform closely follows the principle of QbD. However, the systematic approach begins with the definition of a standard target process. This method is subsequently used for all next-in-class molecules.

Thereafter, the identification of all process parameters starts and are analyzed using a risk-based approach on their impact on the Quality Attributes of the molecule.

During the implementation of the platform technologies, the knowledge space is explored depending on the available prior knowledge. Additional experiments are necessary to acquire an increased number of products and processes to define a Design and Control Space. For various molecules once the platform process is established and successfully demonstrated to be successful, no additional formulation or process development work is required. Whether the lead molecule can be considered a next-in-class molecule could be justified by the information achieved in late discovery and this will help know the Design and Control Space of the established platform.

Initial activities should focus on the basic characterization and formulation and process development of a given class of molecules about which prior knowledge is limited. Extensive studies during initial screening will provide information about suitable formulations. To simplify the process transfer activity a standardized manufacturing process is performed to assess the robustness rather than efficiency and a lead formulation is selected from the result of the above-mentioned stability assessment. The selected lead formulation can be used for subsequent clinical supply manufacturing. Additional backup formulations of any new molecule of the same class will be selected for the test.

In the next phase, the previously identified formulation can be used to confirm whether the molecules y- and z-mAb of the same class exhibit comparable characteristics (next-in-class molecules) or not.

These molecules are characterized in detail and their stability in the lead and in the backup formulations are analyzed. If the data for y- and z-mAb are found to be stableas x-mAb, the standardized formulation and manufacturing process can be used as a platform for this class of molecules. Because of extensive screening of preformulation candidates and the application of prior knowledge from x-, y- and z-mAb, any following next-in-class molecule such as next-mAb can be directly submitted for clinical supply manufacturing. Subsequent IND stability studies will ultimately verify the use of the platform process.

Recent advancements in protein engineering provide the tools to identify and create many potential antibody candidates toward a given target. A systematic screening method can assess whether the molecules possess drug-like properties and can identify the most promising molecules that match predefined stability criteria, as specified in the target molecule profile. In fact, the molecules matching with these criteria would show comparable characteristics and become next-in-class molecules and these can be subsequently developed using a technology platform using a standardized formulation such as composition, excipients, and packaging components, standardized process equipment and unit operations.

Use the prior knowledge can significantly reduce the development efforts required to start phase I clinical trials; while understanding of continuous improvement of the product and process with every molecule of the respective class becomes is very beneficial. Furthermore, the increased predictability and the accuracy of in-time development with synchronized multiple projects are the additional benefits.

A. Multiple Choice Questions

1. Which of the following statements is correct?
    a) Absorption of a drug depends only on the solubility of the drug
    b) Absorption of a drug depends only on the partition coefficient of the drug.
    c) Absorption of a drug depends only on the permeability of the drug
    d) Absorption of a drug depends only on the solubility, partition coefficient, and permeability of the drug

2. During the pilot plant scale production, which of the following information is not documented?
    a) The amount of drug and excipients purchased
    b) Type of personnel required for production
    c) Storage conditions for the intermediates and final product
    d) Shelf-life of the product

3. If 1000 units of a product are manufactured in the laboratory, the lot size on pilot scale would be
    a) 5000   b) 10000
    c) 100000   d) 1000000

4. The specifications for raw materials are developed finally in
    a) Laboratory manufacturing   b) Pilot scale manufacturing
    c) Scale up manufacturing   d) Commercial manufacturing

5. Which of the following is finalised during pilot scale manufacturing?
    a) Time required for completion of each and all the steps
    b) Master formula for manufacturing the dosage form
    c) Guidelines for manufacture and process control
    d) List of the equipment required for manufacturing

6. What is the objective of scale up technique?
    a) Environmental conditions required for each process
    b) Identification of the critical steps and parameters involved in each process
    c) Type and quality of packaging materials required
    d) To determine the processing loss.

7. What of the following points were not considered before the purchase of the equipment?
    a) Cost of the equipment
    b) How frequently the equipment is used?
    c) What would be processing loss?
    d) What is the resale value of the equipment?

8. Irrespective of the method used to manufacture of tablet, which of the following steps is common?
    a) Material handling   b) Drying
    c) Wet granulation   d) Crushing

9. Which of the following parameters is not optimized to optimize the process of blending?
    a) Time of blending   b) Mechanism of blending
    c) Size of blender   d) Blender loading

10. An effective granulation cannot optimize the following parameter?
    a) Composition of the mixture
    b) Concentration of binder
    c) Duration of mixing
    d) Type of mixer used for granulation

11. Which of the following statements is correct?
    a) Bulk density of the powder mix can be decreased
    b) True density of the powder is decreased
    c) Granule density of the blended powder is decreased
    d) Bulk density of the powder mix can be increased

12. Compression of granules performs which of the following activities?
    a) Mixing of the powders
    b) Selection of suitable tablet press
    c) Filling of empty die cavity with blended granules
    d) To determine the characteristics of the raw materials

13. The quality and finish of the coat does not depend on
    a) The expertise of the operator
    b) Strength of the tablets
    c) Design of the core tablets
    d) Size of the tablets compressed

14. Which of the following should not be considered a factor for the encapsulation of powders?
    a) True density of the drug
    b) Bulk density of the powder
    c) Filling volume of the capsule
    d) Lubrication of the powder/granules

15. The viscosity of a dispersion, η, can be expressed as
    a) $\eta = \eta_o + \varphi$ b) $\eta = \eta_o(1 + \varphi)$
    c) $\eta = \eta_o(1 + 2.5\ \varphi)$ d) $\eta = \eta_o \times \varphi$
    (Where, η is the viscosity of the dispersion, η_o is the viscosity of the dispersion medium, and φ is the volume fraction of the phase.)

16. Which of the following parameters does not belong to the processing of the semisolid dosage form?
    a) Temperature and humidity of the manufacturing area
    b) Temperature of the raw materials at which emulsification to be performed
    c) Rate of cooling of the product
    d) Temperature of the aqueous and oil phases

17. Which of the following information is not related to dosage form-manufacturing method?

    a) Information related to the synthetic routes and purification methods.

    b) Information related to crystallization, salt and properties of powders.

    c) Information related to operating conditions.

    d) Information related to environmental load and process safety.

18. Which of the following is related to the history of development of a drug product?

    a) Critical facilities for final processing that may affect the physicochemical properties

    b) Investigation report on causes of abnormalities

    c) Manufacturing records of investigational drugs or samples

    d) Specifications and test methods of intermediates and final drug product

19. Which of the following is not considered SUPAC guidelines?

    a) Information related to changes in the composition and formulation of the quality of finished products

    b) Information related to changes in process variables on the quality of finished products

    c) Information related to monitoring of temperature for drug substances, which require cold storage

    d) Information related to changes in the process site on the quality of the finished products.

20. To establish and justify the applicability of a Platform Technology, it is necessary

    a) To change the number of excipients of total of unit product weight

    b) To identify the molecules which can be considered the next-in-class molecule?

    c) To provide the information to ensure the product quality and performance characteristics of a modified release or immediate release solid oral dose formulation for specified post approval changes.

    d) None of the above

B. Short Questions

1. What is the information about a drug are to be known before selection and development of an appropriate dosage form?

2. During pilot plant scale production of a dosage form, what information is to be documented?

3. Write down the reasons for which the scale up technique is required in the pharmaceutical industry.

4. Briefly mention the reason why the knowledge on 'production rate' in the pharmaceutical industry is necessary.

5. What are the reasons for which granulation is done either by the dry or wet granulation method for making tablet?

6. Enlist the information related to raw materials, intermediates and drug substances required to be documented before the transfer of the technology in a pharmaceutical industry.

7. What are the areas or items required to be verified for reviewing scale-up of a reaction process?

8. Mention in short, the primary aspects to be discussed and defined in SUPAC guidelines.

9. In Platform technology, what is meant by 'Target Molecular Profile'?

C. Long Questions

1. Explain the significance of personnel requirement in pharmaceutical industries.

2. Write a brief note on the four types of spaces required in pilot plant of pharmaceutical industries.

3. Explain the importance of 'Material Handling System' in pharmaceutical industry with respect to pilot plant manufacturing.

4. Explain, why and how the granulation step is important in pilot plant manufacture in pharmaceutical industry?

5. Briefly discuss the most important scale-up parameters related to the manufacture of semisolid dosage forms.

6. Write down the information related to raw materials, intermediates, drug substances, and manufacturing methods are collected during R&D processes before the transfer of technology.

7. Write a brief note on SUPAC guidelines with respect to pharmaceutical product manufacturing.

8. Explain in brief the prerequisites to establish a technology platform for early development.

9. Discuss in brief how the standardized screening process be used to assess and identify next-in-class molecules as defined by the Target Molecular Profile (TMP).

10. Explain, how a platform process can be established in the early stage of development of a drug product?

MCQs Answers

1	2	3	4	5	6	7	8	9	10	11	12	13	14	15	16	17	18	19	20
D	A	B	C	A	B	D	C	B	A	D	C	D	A	C	A	B	D	C	B

CHAPTER 2

Technology Development and Transfer

- ➢ WHO Guidelines for Technology Transfer Technologies
- ➢ Technology Transfer Protocol
- ➢ Technology Transfer Process
- ➢ Quality Risk Management
- ➢ Technology Transfer/Transfer of Technology (TOT)
- ➢ Transfer of Technology (TOT) Agencies in India
- ➢ Asian and Pacific Centre for Transfer of Technology (APCTT)
- ➢ National Research Development Corporation (NRDC)
- ➢ Technology Information, Forecasting and Assessment Council (TIFAC)
- ➢ Biotech Consortium India Limited (BCIL)
- ➢ Confidentiality Agreements
- ➢ Licensing, Technology of Transfer (TOT)
- ➢ Legal Issues on Technology Transfer

Introduction

Technology transfer is the process of transferring scientific findings, knowledge, manufacturing process, technologies, processes, etc. From one organization to another for further development and commercialization. This process typically includes

a) Identifying and assignment of new technologies

b) Technology evaluation

c) Protecting technologies

d) Value addition

e) Marketing and licensing

The main aim of technology transfer is to develop inventions from the lab into marketable products so that the public at large can benefit from the research as quickly and efficiently as possible. Thus, Technology Transfer is the process by which technology is disseminated. It involves communicating relevant knowledge by the transferor to the recipient. It is in the form of a technology transfer transaction which way or may not be a legally binding contract.

There are two terms which are involved in this activity– technology transfer and technology acquisition. The verb "Acquire" means

- To come into possession of getting one's own

- To gain for oneself through one's actions or efforts

Technology Acquisition is the process of acquiring new technology, new product, process, or service; by the efforts of an individual or an enterprise or any other macroentity. This process can be performed either internally or externally within the enterprise.

Types of Technology Transfer

- Scientific Knowledge Transfer, Direct Technology Transfer, Spin-off Technology Transfer

- Informal Technology Transfer and Formal Technology Transfer

- Internal Technology Transfer and External Technology Transfer

Internal Technology Transfer

Internal Technology Transfer refers to such technology transfers/investments where control on the ownership and usage of technology resides with the transferor.

It is a complex process involving the following decisions:

- Timing: When to introduce new technology/products in the market?

- Location: Where to transfer new technology/products?
- Multi-functional teams: Which staff members should be involved in the transfer process?
- Communication method and procedures: What type of communication method and procedures be adopted to facilitate transfer?

Barriers to Internal Technology Transfer

- R and D goals are not known to Production Department.
- Difficulties in stopping current production to test new products/processes
- R&D Department does not understand needs and capability of Production Department.
- Generally, Production Department is innovation-resistant and is bound by routine.
- Non-linkage of new technologies for marketing/customer needs

Overcoming Barriers to Internal Technology Transfer

- Top management support and participation in the transfer process
- Providing a supportive organizational culture
- Use of multi-functional teams in the transfer process
- Ensuring effective communication in the organization
- Bringing R&D closer to production.
- Rotation of a few people between R&D and production
- Linking and participation of marketing elements in the transfer process.

External Technology Transfer

In these transfers, control on the ownership and usage of technology usually does not remain with the transferor, and it passes on to the recipient, as a joint venture with local control, licensing agreement, etc. Successful external technology transfer depends on the following factors: The type of technology being transferred

- The complexity of the technology being transferred
- Transfer mechanism selected
- Relationships between the parties – building of mutual trust
- Core competencies of the parties and compatibility thereof
- The organizational culture of the parties and the mutual understanding thereof

Methods of External Technology Transfer

- Co-operative and collaborative ventures/strategic alliances
- Licensing agreements
- Contracting agreements
- Enterprise acquisition

Why External Technology Transfer?

- The technology already developed saves time and efforts
- Sometimes, growth objectives or competitive goals cannot be reached through internal development
- Lack of risk taking the ability for innovations
- Lack of internal resources (physical and human) for innovation
- The firm does not have core competencies to dealing with complex technological developments.
- Need to keep up with competitors
- We need to cope with the acceleration of technological change
- As a part of firm' strategy—let other firms take big risks and it will purchase technology developed by them.

Barriers to External Technology Transfer

- Associated costs – usually high prices have to be paid in the form of royalties, technical and knowhow fees, etc. over medium to long-term period
- Appropriateness of technology, i.e. Its suitability to core competencies and market needs has always been a point of discussion and investigation
- Heavy reliance on foreign technology- may make transferee/recipient technologically dependent on external technology providers/transferors even for small issues
- Lack of mutual trust between two parties may hinderthe full and timely transfer
- There is a risk of loss of control over technology, and the transferee/recipient may use the technology in an arbitrary manner
- The transfer may render existing technology and its related products/ services/processes obsolete
- This transferee may be a potential competitor in the future.
- Mismatch in the core competencies of the transferor and transferee may create difficulties in transfer

- Different organizationcultures may create difficulties in transfer
- A lack of effective communication botwoon the parties may also create difficulties in transfer

WHO Guidelines for Technology Transfer

Transfer of technology has been defined by WHO as "*a logical procedure that controls the transfer of any process with its documentation and professional expertise between development and manufacture or between manufacture sites.*" It is a systematic procedure that is followed to pass the documented knowledge and experience gained during development and/or commercialization to an appropriate, responsible and authorized party.

Technology transfer represents both the transfer of documentation and the demonstrated ability of the receiving unit (RU) to effectively perform the critical elements of the transferred technology, to the satisfaction of all parties and any applicable regulatory bodies.

Thebusiness strategies of pharmaceutical companies are ever-changing and more, and more involve intra- and intercompany transfers of technology for reasons such as the need for additional capacity, relocation of operations or consolidations and mergers. The WHO Expert Committee on Specificationsfor Pharmaceutical Preparationsrecommended in its forty-second report that WHO report this issue throughthe preparation of WHO guidelines on this matter.

Transfer of technology requires a documented, planned approach using trained and knowledgeable personnel working within a quality system. All aspects of development, production and quality control are covered and documentated. Usually, there are three units:

1. Sending unit (SU),
2. Receiving unit (RU),
3. The unit managing the process, which may or may not be a separate entity.

For "contract manufacturing" and for successful transfer, the following general principles and requirements should be met:

- The project plan should include the quality aspects of the project and e-based on the principles of quality risk management;
- The capabilities of the SU and at the RU should be similar, but not necessarily identical, and facilities and equipment should operate according to the similar operating principles;

- A complete technical gap analysis between the SU and RU including technical risk assessment and potential regulatory gaps should be performed as required;

- Adequately trained staff should be available or should be trained at the RU:

  ➢ Regulatory requirements in the countries of the SU and RU, and in any countries where the product is intended to be supplied, should be taken into account and interpreted constantly throughout any transfer program project; and

  ➢ There should be an effective process and product knowledge transfer.

Technology transfer can be considered successful, if there is documented evidence that the RU can routinely reproduce the transferred product, process, or method against a predefined set of specifications as agreed with the SU.

If the RU identifies particular problems with the process during the transfer, the RU should communicate back to the SU to ensure continuing knowledge management.

Technology transfer projects, particularly those between different companies, have legal and economic implications. If such issues, which may include intellectual property rights, royalties, pricing, conflict of interest and confidentiality, are expected to impact on the open communicationof technical matters in any way, they should be addressed before and during planning and execution of the transfer. Any lack of transparency may lead to the ineffective transfer of technology.

Technology Transfer Protocol

The transfer protocol should enlist chronologically the proposed stages of the transfer. The protocol should include:

- Purposeto provide documented evidence that can be reproduced consistently against a set of predefined specifications (developed by R&D).

- Objective to explain necessary information to transfer technology from R&D to actual manufacturing (commercial manufacturing) by sorting out the information obtained during R&D work.

- The scope of this protocol is applicable to complete technology transfer of products within (Company Name).

- Key personnel and their responsibilities
- A parallel comparison of materials, methods, and equipment;
- The stages of transfer with documented evidence that each critical stage has been satisfactorily completed before the next begins;
- Identification of critical control points;
- Experimental design and acceptance criteria for analytical methods;
- Information about trial production batches, qualification batches, and process validation;
- Change control for any process deviation encountered;
- Evaluation of end-product;
- Arrangements for keeping retain-samples of active ingredients, intermediates, and finished products, and information on reference substances where applicable; and
- Conclusion, including approval duly signed by the project manager.

Names of the personnel	Designation	Responsibilities
	R&D	i. Selection of raw materials ii. Design and development of the manufacturing process for iii. Identification of critical process parameters iv. Establishing specifications and analytical test methods v. Verification of results
	Production	i. Provide facility and equipment for operating ii. Performing operation under control iii. Record results of operation and controls
	Maintenance	i. Calibration and maintenance of equipments
	QC	i. Quality testing of in-process material ii. Quality testing of finished product iii. Preparation of certificate of analysis (CoA)

	Designation	Responsibilities
	QA	i. Preparation and review of documentation (protocol and reports) for all processes of technology transfer
	Head QA	i. Approve the documentation (protocol and reports) for all processes of technology transfer

Technology Transfer Protocol Approval

The technology transfer protocol is actually prepared by a team constituted of various representatives from various disciplines of cross-functional departments; that is, the member experts of the team are from different departments including Research & Development (R&D), Quality Control (QC), Production Department, Quality Assurance (QA) and Maintenance. The constitution of the team should be recorded as per the format shown below:

Name	Designation	Signature and Date

Technology Transfer Process

On the basis of the basic data about the efficacy and safety of the materials obtained from various studies, the quality of the product is designed. Various standards for manufacturing and tests will beestablished in-process while reviewing the production in a manufacturing house and will be subsequently considered to decide the quality of product. This is done when the actual production is started. Technology transfer consists of actions taken in these flows of development to realize the quality as designed during the development stage. The technology transfer processes are classified broadly as mentioned below:

A. **Quality Design (Research Phase)**

During experimental work at the laboratory scale, various experiments are designedto comply with international quality standards and the viability of the product with respect to the yield of the product. The following lab experiments were conducted:

B. **Scale-up and detection of quality variability factors (Development Phase)**

(i) *Research on factory production*

To manufacture as per quality standards obtained during lab experiments, it is required to establish an appropriate quality control method and manufacturing method, after detecting variability factors to secure stable quality in the scale-up batches that are performed to realize factory production of designed by results from small-scale experiments.

Checklist – Laboratory equipment (to be recorded)

Sl. No.	Name of the equipment

Checklist – Raw Material Specifications (to be recorded)

Checklist – Process equipment (to be recorded)

(ii)   *Pilot Batches*

Experimental Data (to be recorded)

Batch No.	Manufacturing Date	Batch size	Yield	Usage of Product	Purity

(iii) *Assurance of consistency through development and manufacturing*

The quality-parameters should be reproducible in the product. For this purpose, Research and Development (R&D) department transfers the developed data to the production department for commercial production.

C.   Technology transfer from R&D Production

R&D department transfers the technology to the production department for commercial production of the product. It is essential to establish the responsibility-system and to prepare the necessary documents to have adequate technology exchange between both departments for successful transfer. While transferring a technology from the R&D department to production department, all relevant technical information must be compiled as a generated report.

D.   Validation and Production (Production Phase)

Based on the studies during conduct of the various processes and at various steps while the manufacturing process is being developed, the Master Manufacturing Formula (MMF) of the product is prepared by R&D department. This is reviewed, finalized and approved by the Quality Assurance (QA) Department by tallying the expected and actually obtained data. From MMF, the Batch Manufacturing Record (BMR) is produced, and production is started in the plant at the commercial scale based on the successful reports obtained during pilot scale batches which are proved to be validated batches. While starting production of a product in the manufacturing plant, the data obtained from the R&D department in the form of MMF is considered and examined carefully Trial Production Batches

E. Experimental data (to be recorded)

Batch No.	Manufacturing date	Batch size	Yield	Usage of product	Purity

F. Benchmarking

Benchmarking is the parameters of measured quality standards of the best market product available. The standards of the in-house products are fixed. Benchmarking provides necessary insights to help us understand how our drug product works comparatively with market samples. In this process of comparison, the data obtained shows how and with which respect our material (product) is comparable to market samples. The HPLC chromatograms were also studied to ascertain the impurity profile.

Summary Table of bench-marking

Sl. No.	Name of sample	Content
	1. In-house sample 2. Market sample - 1	

Affix Chromatogram of Blank, Standard, In-house sample, Market sample – 1, Market sample – 2, etc.

Critical steps and control range of process parameters

Sl. No.	Critical step	Parameter	Control range

Process Specifications

Sl. No.	Test	Specification

Process Flow Charts

Each product is prepared or manufactured through multiple processes. The conditions or parameters of each process accordingly differ and are specific also. For commercial production, such parameters and standards are tried to maintain. If any deviation is found to be compulsory or unavoidable, it must be noted along with the justification. Once the MMF is stabilized and validated, it must be followed regularly. A sample process flowchart in technology transfer is shown below in Fig. 2.1.

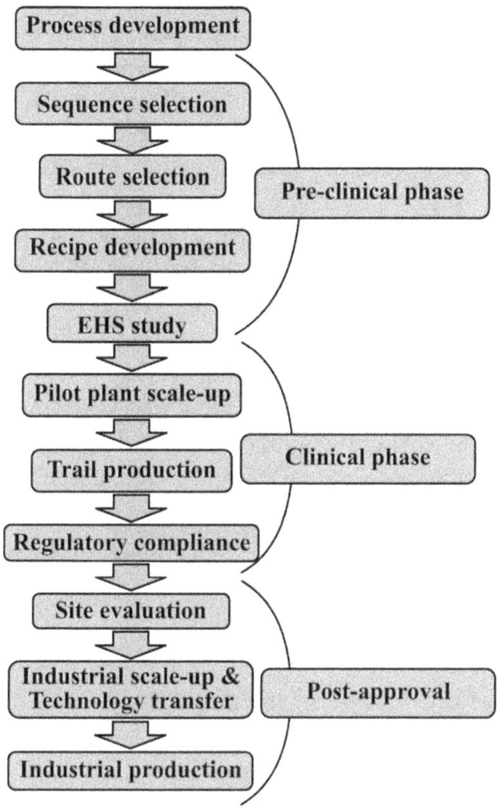

Fig. 2.1 Typical process development steps involved in technology transfer

The document such preparation should be signed by the respective person; hence each document should have the record, 'Prepared & submitted by the following:

Quality Risk Management

WHO Guidelines on Quality Risk Management

In most countries, the compliance with good manufacturing practices (GMP) (including validation), medicines regulatory activities and inspections, with supply chain controls throughout the product life-cycle, are maintained. These provide good quality assurance; in other words, the risks are largely controlled. Quality risk management (QRM) is a process that is relevant to all countries and should provide a rationale for understanding risk and lessen it through appropriate and robust controls. These guidelines aimto assist the development and implementation of effective quality risk management (QRM), covering the activities such as research and development, sourcing of materials, manufacturing, packaging, testing, storage, and distribution. In pharmaceutical industries, hazard analysis and critical control point (HACCP) methodology, traditional food safety management systems have been followed. Subsequently, other industries started to follow. A protocol for risk management has been formulated on the basis of WHO risk management guidance to the pharmaceutical industry.

Principles of Quality Risk Management

It is neither always appropriate, nor always necessary to use a formal risk management process (using recognized tools and internal procedures, such as 'standard operating procedures (SOPs)'. The use of an informal risk management process (using empirical tools or internal procedures) can also be considered acceptable. The two primary principles of QRM are that:

- The evaluation of the risk to quality should be based on scientific knowledge and ultimately linked to the safety of the patient.

- The level of effort, formality, and documentation of the QRM process should be proportionate to the level of risk.

- In addition to these two principles mentioned above, the following principles are also considered part of the QRM methodology:

- When applied to the processes the QRM method should be found to be dynamic, iterative and responsive to record even if, there is a minor change or deviation.

- The capability for continuous improvement should be fixedwithin the QRM process.

This guidance describes the WHO approach to QRM, using the concepts described in ICH Q9. The importance of each component of the framework might differ from case to case, but must be a robust process. It shouldinclude consideration of all elements at a particular level that is proportionateto the specific risk.

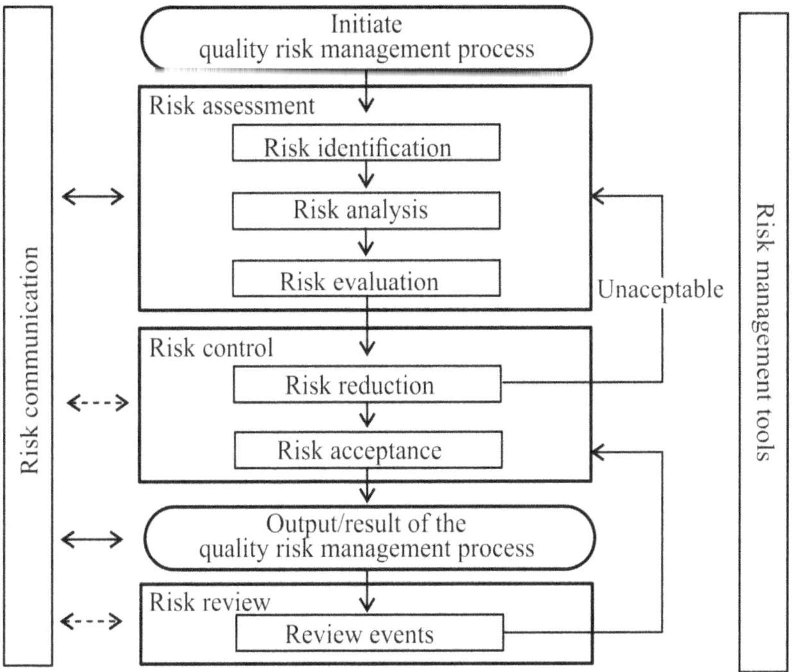

Fig. 2.2 Overview of typical quality risk management processes

General Quality Risk Management Process

Quality risk management is a systematicapproach for the assessment, control, communication and review of risks to the quality of a drug productacross the product lifecycle. A model for quality risk management is outlined in figure 2.2. However, other models can be used. The importance of each component of the framework might differ from case to case, but as a whole, the process should exhibit robustness and the process should incorporateconsideration of all elements.

Responsibilities

Quality risk management activities are usually, but not always, undertaken by interdisciplinary teams. When teams are formed, the members selected should be expert from the appropriate areas, such as, quality control, quality assurance, business development, engineering, regulatory affairs, production operations, sales and marketing, legal, statistics and clinical departments, in addition to individuals who are knowledgeable about the quality risk management process. The following points should be considered during decision making:

- Ability to take responsibility for coordinating quality risk management across various functions and departments of their organization; and

- Given ensured that a quality risk management process is defined, deployed and reviewed, so that adequate resources are available.

Initiating a Quality Risk Management Process

Quality risk management should include systematic processes which are designed to coordinate, facilitate and improve science-based decision making related to risk. Possible steps used to initiate and plan a quality risk management process that should include the following:

- Define the problem and risk in question, including relevant to assumptions for identifying the potential risk;

- Assemble the background information and data on the potential hazard, risk/harm to human health, impact of the human health hazards relevant to the risk assessment;

- Identify a leader and the necessary resources;

- Specify a timeline within which the process can be completed or delivered, and the appropriate level of decision to be made for the risk management process.

Risk Assessment

Risk assessment consists of (1) the identification of hazards, (2) the analysis and (3) evaluation of risks associated with exposure to those hazards (as defined below). Quality risk assessments start with a well-defined problem description or risks in question. When the risk in question is well defined, an appropriate risk managementtool and the types of information required to be addressedand therisk in question will be more readily identifiable. As a support to clearly define the risk(s) for risk assessment purposes, answers to three fundamental questions are often helpful:

1. What might go wrong?

2. What is the likelihood (probability) that it will go wrong?

3. What are the consequences (severity)?

Risk identification is a systematic use of information to identify hazards referringto the risk in question or problem description. Information can include historical data, theoretical analysis, informed opinions, and the concerns of stake-holders. Riskidentification clarifies "What might go wrong?" Including "identification of possible consequences." This provides the basis for further steps in the quality riskmanagement process.

Risk analysis is the estimation of the risk related to the identified hazards. It is the qualitative or quantitative process of linking the probability of occurrence andseverity the of harms/risks. In some risk management tools, the ability to detect the harm (detectability) is also provided or assessed. The factors in the estimation of risk should be considered.

Risk evaluation compares the identified and analyzed risk against given criteria for the risk. Risk evaluations consider the strength of evidence for all three fundamental questions mentioned above.

In doing an effective risk assessment, the robustness of the dataset is important because it determines the quality of the output. Revealing assumptions and reasonable sources of uncertainty will improve confidence in this output and help to identify its limitations. Uncertainty is due to a combination of incomplete knowledge about a process and its expected or unexpected variability. Typical sources of uncertainty include gaps in knowledge, gaps in pharmaceutical science, and process understanding, sources of harm, such as, failure modes of a process, source of variability, and the probability of detection of problems.

The output of a risk assessment is either a quantitative estimate of risk or a qualitative description of the range of risk. A numerical probability is used to express the risk quantitatively. Alternatively, risk can be expressed qualitatively by using descriptors, such as "high," "medium," or "low," which should be defined in as muchdetail as possible. Sometimes a "risk score" is used to further define descriptors in riskranking. In quantitative risk assessments, a risk estimate provides the probability of aspecific consequence, under a given a set of risk-generating circumstances. Thus, quantitative risk measurement is useful for one particular consequence at a time. Alternatively, somerisk management tools use a relative risk measure to combine multiple levels ofseverity and probability into an overall estimate of relative risk. The intermediatesteps within a scoring process can sometimes use quantitative risk estimation.

Risk control includes decision making to reduce and accept risks. The purpose of risk control is to reduce the risk to an acceptable level. The amount of effort used forcontrolling the risk should be proportional to the significance of the risk. Decision-makers can use different processes. The process includes the cost- benefit analysis, for understanding theoptimal level of risk involved. Risk control might focus on the following questions:

- Is the risk above an acceptable level?
- What can be done to reduce or eliminate these risks?
- What is the appropriate balance among benefits, risks, and resources?
- Are new risks introduced because of the identified risks being controlled?

Risk Reduction

It focuses on processes for mitigation or avoidance of quality risk whenit exceeds a specified (acceptable) level (see Fig. 2.1). Risk reduction might includeactions taken to mitigate the severity and the probability of harm. Processes that improvethe detectability of hazards and quality risks might also be used as part of a riskcontrol strategy. The implementation of risk reduction measures can introduce newrisks into the system or increase the significance of other existing risks. Therefore, itmight be appropriate to revisit the risk assessment to identify and evaluate anypossible change in risk after implementing a risk reduction process.

Risk Acceptance

Itis a decision to accept the risk. Risk acceptance can be a formaldecision to accept the residual risk, or it can be a passive decision in which residualrisks are not specified. For some types of ills, even the best quality riskmanagement practices might not eliminate the risk. In these circumstances, itmight be agreed that an appropriate quality risk management strategy has beenapplied and that quality risk is reduced to a specified (acceptable) level. This (specified) acceptable level will depend on many parameters and should be decided by one case-by-case basis.

Risk Communication

Risk communication is the sharing of information about the risk and management of riskbetween the decision-makers and others. Parties can communicate at any stage of therisk management process (Fig. 2.1: dashed arrows). The output/result of the quality risk management process should be appropriately communicated and documented (Fig. 2.1: solid arrows). Communications may include those among interested parties, such as, regulators and industry, industry and the patient, within a company, industry or regulatory authority, etc. The included information may relate to theexistence, nature, form, probability, severity, aooeptability, control, treatment, detectability or other aspects of risks to quality. Communication need not be carriedout for every risk acceptance. Between the industry and regulatoryauthorities, communication regarding the quality risk management decisions may be through existing channels as specified in regulations and guidance.

Risk Review

Risk management should be an ongoing part of the quality management process. A mechanism to review or monitor the events should be executed. The output/results of the risk management process should be reviewed to take into account new knowledge and experience. Once a quality risk management process has been initiated, that process should continue to be used for events that may influence the original quality risk management decision, whether these events areplanned, such as, results of product review, inspections, audits, change control, orunplanned, such as, the cause of failure in investigations, recall. The frequency of any review should be based on the level of risk. Risk review may include reconsideration the of risk acceptance decisions.

Quality Risk Management Methodology

Quality risk management supports a scientific and practical approach to decision-making. It provides documented, transparent and reproducible methods to completethe steps of the quality risk management processes based on the current knowledge about assessing the probability, severity and sometimes delectability of the risk.

Traditionally, risks to quality have been assessed and managed in various informal ways (empirical and internal procedures) by a compilation of observations, trends, and other information. Such approaches continue to provide useful information that may support the topics such as handling of complaints, quality defects, deviations and allocation of resources. Additionally, the pharmaceutical industry and regulatory authorities can assess and manage the riskusing recognized risk management tools and internal procedures such as, standard operating procedures (SOPs). A non-exhaustive list of some of these tools is given below:

- Basic risk management facilitation methods (flow-charts, check-sheets, etc.);
- Failure Mode Effects Analysis (FMEA);
- Failure Mode, Effects and Criticality Analysis (FMECA);

- Fault Tree Analysis (FTA);
- Hazard Analysis and Critical Control Points (HACCP);
- Hazard Operability Analysis (HAZOP);
- Preliminary Hazard Analysis (PHA);
- Risk ranking and filtering;
- Supporting statistical tools

It may be appropriate to adapt these tools for use in specific areas aboutdrug substance and drug product quality. Quality risk managementmethods and the supporting statistical tools can be used in combination; for example, Probabilistic Risk Assessment. Combined-use provides the flexibility that can facilitatethe application of quality risk management principles.

The degree of precision and formality of quality risk management should reflect the availableknowledge and may correspondingto the complexity and criticality of the issue tobe addressed.

Technology Transfer Process

The quality of a drug is designed on the basis of primary data related to efficacy and safety obtained from various studies in preclinical phases and data relating to the efficacy, safety, and stability of drug products obtained from clinical studies. The quality of design will be almost completed in a Phase II clinical study. Various standards for manufacturing and analysis will be established in the process of reviewing the production in the factory and Phase III study to realize the quality of the design. The quality of the design will be verified in various validation studies and will be upgraded to be the quality of the product, and the actual production that will be started. The technology transfer consists of actions taken in these flows of development to realize the quality as designed during the manufacture. Even if the production starts, the technology transfer will occur in processes such as changes in manufacturing places. These processes are classified broadly into the following five categories.

1. Quality Design (Research Phase)
2. Scale-up and Detection of Quality Variability Factors (Development Phase)
3. Technology Transfer from R&D to Production
4. Validation and Production (Production Phase)
5. Feedback of Information Generated from the Production Phase and Technology Transfer of Marketed Products

Organization and Management

- Transfer of technology covers a sending unit (SU) and a receiving unit (RU). In some cases, there may be an additional unit, which would be responsible for directing, managing and approving the transfer.

- There is a formal agreement between the parties, which specifies the responsibilities before, during and after transfer.

- Organization and the management of a successful technology transfer requires ensure that the main steps have been executed and documented.

- The specifications and relevant functional characteristics of the starting materials (APIs and excipients) to be used at the RU should be consistent with materials used at the SU. Any properties, which are likely to influence the process or product should be identified and characterized.

Active Pharmaceutical Ingredients (API)

The SU should provide the RU with the applicant's API master file (APIMF) or drug master file (DMF) or active substance master file (ASMF)), or equivalent information and any relevant additional information on the API of importance for the manufacture of the pharmaceutical product.

The following are the examples of the information which may typically be provided; however, the information needed in eachspecific case should be assessed using the principles of QRM:

- Manufacturer and associated supply chain;

- The step of the API to be transferred;

- Flowchart of synthesis pathway, outlining the process, including entry points for raw materials, critical steps, process controls and intermediates;

- Where relevant, a definitive physical form of the API (including photomicrographs and other relevant data) and any polymorphic and solvate forms;

- Solubility profile;

- If relevant, pH of solution;

- Partition coefficient, including the method of determination;

- Intrinsic dissolution rate, including the method of determination;

- Particle size and size-distribution, including the method of determination;

- Bulk physical properties, including data on bulk and tap density, surface area and porosity as appropriate;

- Water content and determination of hygroscopicity, including water activity data and special handling requirements;

- Microbiological considerations (including sterility, bacterial endotoxins and bioburden levels where the API supports microbiological growth) by national, regional, or international Pharmacopoeial requirements;

- Specifications and justification for release and end-of-life limits;

- Summary of stability studies conducted in conformity with current guidelines, including conclusions and recommendations on retest date;

- List of potential and observed synthetic impurities, with data to support the proposed specifications and typically observed levels;

- Information on degradants, with a list of potential and observed degradation products and data to support the proposed specifications and typically observed levels;

- Potency factor, indicating observed purity and justification for any recommended adjustment to the input quantity of API for product manufacturing, providing example calculations; and

- Special considerations with implications for storage and or handling, including but not limited to safety and environmental factors (e.g. As specified in material safety data sheets) and sensitivity to heat, light, or moisture.

Considerations

The SU should provide any information on the history of the process development which may be required to enable the RU to perform any further development and/or process optimization after a successful transfer. Such information may include the **EXCIPIENTS**

The excipients used have a potential impact on the final product. Their specifications and relevant functional characteristics should, therefore, be made available by the SU for transfer to the RU site. The following are examples of the information which may typically be provided; however, the information needed in each specific case should be assessed using the principles of QRM:

- Manufacturer and associated supply chain;

- Description of functionality, with justification for inclusion of any antioxidant, preservative, or any excipient;
- Definitive form (particularly for solid and inhaled dosage forms);
- Solubility profile (particularly for inhaled and transdermal dosage forms);
- Partition coefficient, including the method of determination (for transdermal dosage forms);
- Intrinsic dissolution rate, including the method of determination (for transdermal dosage forms);
- Particle size and distribution, including the method of determination (for solid, inhaled and transdermal dosage forms);
- Bulk physical properties, including data on bulk and tap density, surface area and porosity as appropriate (for solid and inhaled dosage forms);
- Compaction properties (for solid dosage forms);
- Melting point range (for semi-solid or topical dosage forms);
- pH range (for parenteral, semi-solid or topical, liquid and transdermal dosage forms);
- Ionic strength (for parenteral dosage forms);
- The specific density or gravity (for parenteral, semi-solid or topical, liquid and transdermal dosage forms);
- Viscosity and or viscoelasticity (for parenteral, semi-solid or topical, liquid and transdermal dosage forms);
- Osmolarity (for parenteral dosage forms);
- Water content and determination of hygroscopicity, including water activity data and special handling requirements (for solid and inhaled dosage forms);
- Moisture content range (for parenteral, semisolid or topical, liquid and transdermal dosage forms);
- Microbiological considerations (including sterility, bacterial endotoxins and bioburden levels where the excipient supports microbiological growth) by national, regional, or international Pharmacopeial requirements, as applicable (for general and specific monographs);
- Specifications and justification for release and end-of-life limits;

- Information on adhesives supporting compliance with peel, shear and adhesion design criteria (for transdermal dosage forms);

- Special considerations with implications for storage and or handling, including but not limited to safety and environmental factors, such as, as specified in material safety data sheets (MSDS) and sensitivity to heat, light, or moisture; and

- Regulatory considerations, such as, documentation to support compliance with transmissible animal spongiform encephalopathy certification requirements (where applicable).

Information on Process and Finished Pharmaceutical Products

The SU should provide a detailed characterization report of the product, including its

1. Qualitative and quantitative composition,
2. Physical description,
3. A method of manufacture,
4. In-process control methods and specifications,
5. Packaging components and configurations, and
6. Any safety and handling, for following:
   - Information on clinical development, e.g., information on the rationale for the synthesis, route and form selection, technology selection, equipment, clinical tests, and product composition;
   - Information on scale-up activities: process optimization, statistical optimization of critical process parameters, critical quality attributes, pilot reports and or information on pilot-scale development activities indicating the number and disposition of batches manufactured;
   - Information or reports on full-scale development activities, indicating the number and disposition of batches manufactured, and deviation and change control (sometimes referred to as change management) reports, which led to the current manufacturing process;
   - The change history and reasons, e.g., a change control log, indicating any changes to the process or primary packaging or analytical methods as a part of process optimization or improvement;
   - Information on investigations of problems and the outcomes of the investigations.

The SU should provide the information on any health, safety and environmental issues associated with the manufacturing processes to be transferred, and the implications, e.g. The need for owning or protective clothing to the RU.

The SU should provide the RU information on current processing and testing, including but not limited to:

- A detailed description of facility requirements and equipment;

- Information on starting materials, applicable MSDS and storage requirements for raw materials and finished products;

- Description of manufacturing steps (narrative and process maps or flow-charts, and/or master batch records), including qualification of processing hold times and conditions, order and method of raw material in addition to the bulk transfers between processing steps;

- Description of analytical methods;

- Identification and justification of a control strategy (e.g., identification of critical performance aspects for specific dosage forms, identification of process control points, product quality attributes and qualification of critical processing parameter ranges, statistical process control (SPC) charts);

- Design space, in cases where this has been defined;

- Validation information, e.g., validation plans and reports;

- Annual product quality reviews;

- Stability information;

- An authorized set of protocols and work instructions for manufacturing; and

- Environmental conditions or any special requirement desired for the facility or equipment, depending on the nature of the product to be transferred

During the technology transfer process, the RU should identify any differences in facilities, systems, and capabilities and communicate with the SU about these differences to understand the potential impact on the ability to run the process to deliver good product quality. Differences should be understood and satisfactorily addressed to ensure equivalent product quality. Based on the information received from the SU, the RU should consider its capability to manufacture and pack the product to the required standards and should develop relevant plant operating procedures and documentation before the start of production. Process development at the RU should address the following tasks:

- Comparison and assessment of suitability and qualification of facility and equipment;

- Description of the manufacturing process and flow of personnel and of materials at the RU (narrative and or process maps or flow charts);

- Determination of critical steps in manufacture, including hold times, endpoints, sampling points and sampling techniques;

- Writing and approval of SOPs for all production operations (e.g. Dispensing, granulation or blending or solution preparation, tablet compression, tablet coating, encapsulation, liquid filling, primary and secondary packaging, and in-process quality control), packaging, cleaning, testing, and storage.

- Evaluation of stability information, with the generation of site-specific stability data if required;

- Compliance with regulatory requirements for any changes made, e.g., Regarding batch size.

Packaging

- The transfer of packaging operations should follow the same procedural patterns as those of the production transfer.

- Information on packaging to be transferred from the SU to the RU includes specifications for a suitable container or closure system, as well as any relevant additional information on design, packing, processing, or labeling requirements and tamper-evident and anti-counterfeiting measures needed for qualification of packaging components at the RU.

- For QC testing of packaging components, specifications should be provided for drawings, artwork and material (for example, glass, card or fiber board).

- Based on the information provided, the RU should perform a suitability study for initial qualification of the packaging components. Packaging is considered suitable if it provides adequate protection (preventing degradation of the medicine due to environmental influences), safety (absence of undesirable substances released into the product), compatibility (absence of interaction possibly affecting medicine quality) and performance (functionality in terms of drug delivery).

Premises

- The SU should provide information to the RU on the layout, construction and finish of buildings and services (heating, ventilation and air-conditioning (HVAC), temperature, relative humidity, water, power, and compressed air), which impact the product, process, or method to be transferred.

- The SU should provide information on relevant health, safety and environmental issues, including

  - Inherent risks of the manufacturing processes (e.g. reactive chemical hazards, exposure limits, fire and explosion risks);

  - Health and safety requirements to minimize operator exposure (e.g. atmospheric containment of pharmaceutical dust);

  - Emergency planning considerations (e.g. in case of gas or dust release, spillage, fire and firewater run-off); and

  - Identification of waste streams and provisions for re-use, recycling and/or disposal.

Equipment

- The SU should provide a list of equipment, makes and models involved in the manufacture, filling, packing and or control of the product, process, or method to be transferred, with existing qualification and validation documentation. Relevant documentation may include:

  - Drawings;

  - Manuals;

  - Maintenance logs;

  - Calibration logs; and

  - Procedures (e.g. Regarding equipment set-up, operation, cleaning, maintenance, calibration, and storage).

- The RU should review the information provided by the SU together with its own inventory list, including the qualification status (IQ, OQ, PQ) of all equipment and systems, and perform a side-by-side comparison of equipment at the two sites in terms of their functionality, makes, models and qualification status.

- The RU should perform a gap analysis to identify requirements for adaptation of existing equipment, or acquisition of new equipment, or a change in the process, to enable the RU to reproduce the process being transferred. GMP requirements should be satisfied and intended production volumes and batch sizes (e.g. The same, scaled-up or campaign) should be considered. Factors to be compared include

➢ Minimum and maximum capacity;

➢ Material of construction;

➢ Critical operating parameters;

➢ Critical equipment components (e.g. filters, screens, and temperature/pressure sensors);

➢ Critical quality attribute; and

➢ Range of intended use.

- The facility- and building-specification of all equipment at the RU should be considered at the time of drawing up process maps or flow charts of the manufacturing process to be transferred, including flows of personnel and material.

- The impact of manufacturing new products on products currently manufactured with the same equipment should be determined.

- Any modification of existing equipment that needs to be adapted to become capable of reproducing the process being transferred should be documented in the transfer project plan.

Documentation

➢ The documentation required by the transfer project itself is wide ranging. Examples of documentation commonly required are summarized in Table 2.

➢ The documented evidence that the transfer of technology has been considered successful should be formalized and stated in a technology transfer summary report. That report should summarize the scope of the transfer, the critical parameters as obtained in the SU and RU (preferably in a tabulated format) and the final conclusions of the transfer. Possible discrepancies should be listed and appropriate actions, where needed, taken to resolve them. DQ, design qualification; IQ, installation qualification; OQ, operational qualification; API, active pharmaceutical ingredient; SOPs, standard operating procedures; RU, receiving unit.

Table 2.1 Some documentation for transfer of technology (TOT)

Key task	Documentation provided by SU	Transfer documentation
Project definition	Project plan and quality plan (where separate documents), protocol, risk assessments, gap analysis	Project implementation plan **TOT protocol**
Quality agreement		
Facility assessment	Plans and layout of facility, buildings (construction, finish) Qualification status (DQ, IQ, OQ) and reports	Side-by-side comparison with RU facilities and buildings; gap analysis Qualification protocol and report
Health and Safety assessment	Product-specific waste management plans Contingency plans	
Skill set analysis and training	SOPs and training documentation (product-specific operations, analysis, testing)	Training protocols, assessment results
Analytical method transfer	Analytical method specifications and validation, including in-process quality control	
Starting material evaluation	Specifications and additional information on APIs, excipients.	-
Equipment selection and transfer	Inventory list of all equipment and systems, including makes, models, qualification status (IQ, OQ, PQ) Drawings, manuals, logs, SOPs (e.g. set-up, operation, cleaning, maintenance, calibration, storage)	-
Process transfer: manufacturing and packaging	Reference batches (clinical, dossier, bio-batches) Development report (manufacturing process rationale) History of critical analytical data Rationale for specifications Change control documentation Critical manufacturing process parameters Process validation reports Drug master file API validation status and report(s) Product stability data Current master batch manufacturing and packaging records List of all batches produced Deviation reports Investigations, complaints, recalls Annual product review	-
Cleaning	Cleaning validation, including Solubility information; therapeutic doses; category (toxicology); existing cleaning SOPs; validation reports — chemical and micro; agents used; recovery study	-

Qualification and Validation

The extent of qualification and/or validation to be performed should be determined based on risk management principles. The principles of risk management are effectively used in different areas of business and government activities, such as finance, insurance, occupational safety, public health, pharmacovigilance, and regulation authorities regulating these industries. Presently, there are examples of using quality risk management in the pharmaceutical industry. The appropriate use of quality risk management can facilitate but does not obviate, the obligation of industry to comply with the regulatory requirements and does not replace appropriate communications between industry and regulators.

The WHO prepares the draft guidelines for validation at different periods for pharmaceutical industries. Validation is a primary part of good practices, such as good manufacturing practices. Therefore, it is an element of the pharmaceutical quality system. As a concept, validation includes qualification during the lifecycle of the product, process, equipment, or utility. These guidelines include the general principles of validation and qualification. The principles that apply are

- The validation should be executed in compliance with regulatory expectations.

- The quality, safety and efficacy must be designed and built into the product.

- The quality cannot be inspected or tested in the product.

- Quality risk management principles should be applied in determining the need, scope and extent of validation.

- An ongoing review should occur to ensure that the validated state is maintained and opportunity for identified continuous improvement.

These guidelines aim mainly at the overall concept of validation. This serves as general guidance only and the principles may be considered useful for its application in the manufacture and control of materials used and finished products. Validation of specific processes and systems such as manufacturing of sterile products, requires much more consideration.

Calibration refers to the set of operations that establish, under specified conditions, the relationship between values indicated by an instrument or system for measuring such as weight, temperature and pH, recording and controlling, or the values represented by a material measure, and corresponding known values of a reference standard. Limits for acceptance of the results of measuring should be established.

Cleaning validation is the documented evidence to establish that cleaning procedures include removal of residues to predetermined levels of acceptability, taking into consideration factors such as batch size, dosing, toxicology and size of equipment.

Commissioning of equipment or instrument refers to setting up, adjustment and testing of equipment or a system to ensure that it meets all the requirements, as specified in the user requirement specification, and capacities as specified by the designer or developer. Commissioning is carried out before qualification and validation.

Qualification refers to documented evidence that all premises, systems, or equipment all can achieve the predetermined specifications that these are properly installed, and/or work correctly and lead to the expected results. There are four types of qualification.

Design qualification refers to the documented verification that the proposed design of facilities, system and equipment is suitable for the intended purpose.

Installation qualification is the documented verification that the installations such as machines, measuring devices, utilities and manufacturing areas.

Performance qualification refers to the documented verification that the equipment or system performs consistently and reproducibly within defined specifications and parameters in its normal operating environment.

Requalification of utilities, systems and equipment should be maintained in a validated state. Any changes made to these should be managed through the change control procedure. The extent of qualification and validation because of such a change should be determined based on the risk management principles. Requalification should be done based on the identified need. The requalification should be considered on the basis of risk management principles. Factors such as the frequency of use, breakdowns, results of operation, criticality, preventive maintenance, repairs, calibration, verification may be considered.

Requalification should also be considered after cumulative/multiple changes. Changes of equipment, which involve the replacement of equipment on a 'like-for-like' basis will require requalification. The replacement of parts may not require full requalification.

Similar to the qualification, validation can be of different types such as

Cleaning validation is the documented evidence to establish that cleaning procedures remove the residues to predetermined levels of acceptability, and the factors such as batch size, dosing, toxicology and size of equipment are to be considered.

Process validation is the collection and evaluation of data, throughout the product life cycle, which provides documented scientific evidence that a process is capable of consistently delivering quality products.

Prospective validation indicates that validation carried out during the development stage on the basis of a risk analysis of the production process, which is broken down into individual steps; these are then evaluated on the basis of experience to determine whether they may lead to a critical situation.

Revalidation means the system has been in place ensuring that the procedures have been in a validated state; for example, through verification cleaning validation and analytical methods, validation can be identified whether revalidation is required or already revalidated.

Validation Master Plan (VMP) is a high-level document that establishes an umbrella validation plan for the entire project and summarizes the manufacturer's overall philosophy and approach, to be used for establishing performance adequacy. It provides information on the manufacturer's validation work program and defines details of and timescales for the validation work to be performed, including a statement of the responsibilities of those implementing the plan. Therefore, each pharmaceutical company must follow the guidelines of the world health organization for their equipment or process validation. In fact, each company must have an internal protocol for the said programme. However, the said protocol should be amended on a regular basis.

Qualification and Validation Protocols

Every industry should have the qualification and validation protocols describing the qualification and validation to be performed. Protocols should include at least the following significant background information:

➢ The objectives

➢ The site

➢ The responsible personnel

➢ Description of the standard operating procedures (SOPs)

➢ Equipment or instrument to be used

➢ Standards and criteria as appropriate

- ➤ Stage of validation or qualification
- ➤ The processes and/or parameters
- ➤ Sampling, testing and monitoring requirements
- ➤ Stress testing where appropriate
- ➤ Calibration requirements
- ➤ Predetermined acceptance criteria for drawing conclusions
- ➤ Review and interpretation of the results
- ➤ Change control, deviation
- ➤ Achieving and retention.

Transfer of Technology (TOT) Agencies in India

The following section explains about different TOT agencies in India.

Asian and Pacific Centre for Transfer of Technology (APCTT)

Overview

- APCTT is a United Nations Regional Institution under the Economic and Social Commission for Asia and the Pacific (ESCAP).

- The Centre was established in 1977 in Bangalore, India. In 1993, the Centre moved to New Delhi, India. APCTT promotesthe transfer of technology to and from small- and medium-scale enterprises (SMEs) in Asia and the Pacific.

- APCTT implements development projects funded by international donors aimed at strengthening the environment for technology transfer among SMEs in Asia and the Pacific; in this respect, the Centre makes special efforts to encourage more participation of women in the field of technology.

- APCTT undertakes consultancy assignments in various technology transfer related areas (institution building, human resources development, studies, business partnership development).

- The objective of APCTT is to strengthen the technology transfer capabilities in the region and to facilitate import/export of environmentally sound technologies to/from the member countries.

- All member states and associate members of UNESCAP are de facto members of APCTT.

Governance

APCTT is governed by a Governing Council consisting of a representative designated by India's Government, which extends a host facility to the Centre, and at least eight representatives nominated by other and associate members of the United Nations Economic and Social Commission for Asia and the Pacific (ESCAP) elected by the Commission for three years. The Governing Council meets at least once a year to advise on the formulation and implementation of the program of work and reviews the administration and financial status of the Centre. The Executive Secretary of ESCAP submits an annual report, as adopted by the Governing Council, to ESCAP at its annual sessions. APCTT also reports its activities to the Committee on Trade and Investment, one of the nine committees that make up the subsidiary structure of ESCAP. The Advisory Committee of Permanent Representatives and Other Representatives Designated by Members of ESCAP periodically reviews the work of APCTT. The present Governing Council elected for the tenure of three years (2014–2017) comprises the 14 member States, such as India, China, Bangladesh, Sri Lanka, Thailand, Malaysia, etc.

Objectives

The objectives of the Centre are to assist the members and associate members of ESCAP through strengthening their capabilities to develop and manage national innovation systems; develop, transfer, adapt and apply technology; improve the terms of transfer of technology; and identify and promote the development and transfer of technologies relevant to the region.

Functions

The Centre will achieve the above objective by undertaking processes such as

a) Research and analysis of trends, conditions, and opportunities;

b) Advisory services;

c) Dissemination of information and good practices;

d) Networking and partnership with international organizations and key stakeholders;

e) Training of national personnel, particularly national scientists and policy analysts.

Status and Organization

(a) The Centre shall have a Governing Council, which is now referred to as "The Council," a Director and staff.

(b) The Centre is located in New Delhi.

(c) The Centre's activities shall be in line with relevant policy decisions adopted by the General Assembly, the Economic and Social Council and the Commission. The Centre shall be subject to the Financial and Staff Regulations and Rules of the United Nations and the applicable administrative instructions.

National Research Development Corporation (NRDC)

Introduction

The National Research Development Corporation (NRDC) was established in 1953 by India's Government, with the primary objective to promote, develop and commercialize the technologies/know-how/ inventions/patents/processes emanating from various national R&D institutions/Universities and is presently working under the administrative control of the Dept. of Scientific & Industrial Research, Ministry of Science & Technology. During the past six decades of its existence and in pursuance of its corporate goals, NRDC has forged strong links with the scientific and industrial community in India and abroad and developed a wide network of research institutions, academia and industry and made formal arrangements with them for the commercialization of know-how developed in their laboratories and is now recognized as a large repository of a wide range of technologies spread over almost all areas of industries, viz. Agriculture and Agro-processing, Chemicals including Pesticides, Drugs and Pharmaceuticals, Bio Technology, Metallurgy, Electronics and Instrumentation, Building Materials, Mechanical, Electrical and Electronics, etc. It has licensed the indigenous technology to morethan 4800 entrepreneurs and helped establish a large number of small and medium scale industries.

Besides being the torch bearer in the field of technology transfer, NRDC also undertakes several activities under its structured promotional program for the encouragement and advancement of research, promotion of inventions and innovations such as meritorious inventions awards, Techno-Commercial support, technical and financial assistance for IPR Protection, value addition services and support for further development of technologies and much more.

The NRDC has also successfully exported technologies and services to both developed and the developing countries. NRDC isrecognized, particularly in the developing countries, as the source of reliable, appropriate technology, machines, and services, which are typically suitable for these countries.

Outreach Centres

a) Intellectual Property Facilitation Centre (I.P.F.C.)

b) Innovation Facilitation Centre (I.F.C.)

a) Intellectual Property Facilitation Centre (I.P.F.C.)

The Intellectual Property Facilitation Centre – IPFC, is a joint project of NRDC and the Ministry of Micro, Small & Medium Enterprises (M/oMSME or MSMEs), which promotes awareness and adoption of Intellectual Property Rights among entrepreneurs and MSMEs in India while making it accessible to high-quality IP services and resources.

NRDC with its vast experience in technology transfer and IP management proposes to identify and facilitate the development of Innovation Facilitation Centres in the Universities, National Institutes of Technology (NIT's), IIT's, Autonomous Institutions & Academic Institutions across the country where innovation activities would be promoted and encouraged among the faculty, students and researchers and would facilitate the effective management of Intellectual Property, development of an association with manufacturing enterprises in the country and abroad and subsequently the transfer of the IP to the industry and entrepreneurs.

b) Innovation Facilitation Centre (I.F.C.)

NRDC in collaboration with Universities, National Institutes of Technology (NIT's), IIT's, Autonomous Institutions & Academic Institutions in India will set up Innovation Facilitation Centres (NRDC-IFC) on their premises. The network of Corporation with the local, regional and global industries and entrepreneurs would facilitate the faculty, students and researchers of the Universities, National Institutes of Technology (NIT's), IIT's, Autonomous Institutions & Academic Institutions in India to market their innovative research work and reap the benefits of commercial success.

Main Objectives

a) To promote area-specific technologies for the industry by using the R&D capabilities of the host institutions in the region

b) To provide value addition servicesregarding IP management and technology commercializationfor making commercial products from the R&D of institutions in the region

c) To act as a potent resource of IP protected technologies to help the manufacturing sector of the country to develop new products and services based on the innovative technologies available through the Innovation Facilitation Centres

d) To facilitate and createan appropriate environment in the Universities, National Institutes of Technology (NIT's), IIT's, Autonomous Institutions & Academic Institutions in India for promoting innovation processes and providing training and support

e) To play the role of a facilitator to stimulate entrepreneurial activities within the Universities, National Institutes of Technology (NIT's), IIT's, Autonomous Institutions & Academic Institutions in India through the NRDC-IFC

f) To provide Intellectual Property protection and management services and facilitate the technology transfer andcommercialization

g) To develop and strengthen the Industry-Academia partnership for development and effective transfer of know-how from the Universities, National Institutes of Technology (NIT's), IIT's, Autonomous Institutions & Academic Institutions in India to the entrepreneurs and industry.

Technology Information, Forecasting and Assessment Council (TIFAC)

Need for undertaking technology forecasting and assessment studies on a systematic and continuing basis was highlighted in the Government of India's Technology Policy Statement (TPS) of 1983. It further made it mandatory, for the Ministries and Agencies with large investments or a large volume of production to provide a technology forecast covering their requirements over a 10 year or longer period and evolve suitable strategies for development based on priorities.

Subsequently, as per the recommendation of the Technology Policy Implementation Committee (TPIC) in 1985, Cabinet approved the formation of TIFAC in mid-1986 and TIFAC was formed as a registered Society in February 1988 under the Department of Science and Technology as an autonomous bodies. It was mandated to assess the state-of-the-art technology and set directions for future technological development in India in important socio-economic sectors.

As a unique knowledge network institution in India, TIFAC activities encompass a wide array of technology areas and fill a critical gap in the overall S&T system of India. The organization has carried out technology foresight exercise, facilitated and supported technology development; prepared technology-linked business opportunity reports and implemented mission-mode programs.

Mandate

- To obtain from appropriate sources and project the estimates of the nature and quantum of the likely demands of goods and services in various sectors of the economy against 10 and 25 years' time-frames on the basis of a) 'normative', and (b) 'exploratory' approaches and to suggest the direction and extent of technological changes that might be considered necessary to fulfill these demands in the light of the existing or anticipated resource base of the country.

- To prepare Technology Impact Statements, to uncover the likely implications and consequences, both desirable and undesirable, of the existing and newly emerging technologies upon society, indicating to decision-makers, throughthe generation of future-oriented scenarios, their short-term and long-term implications.

- Based on the T.I.F. & A. Studies and with a view to –

  (a) Ensuring timely availability of requisite technologies relevant to the needs of the country on a futuristic basis and minimizing the time gap between the development of new technologies and their usage and

  (b) Establishing a purposeful linkage between technology development and technology import policies, to identify priority areas of research in relation to the socio-economic, environmental, and security needs of the country; to evolve and suggest strategies for technological developments based on such priorities; and to draw up programs of purposeful research in various sectors.

- To fulfill the above objectives, to devise and set up suitable Information Collection, Analysis and Programming groups.

Biotech Consortium India Limited (BCIL)

The BCIL was inaugurated in 1990 by the then Prime Minister of India, Hon'ble Shri Chandra Shekhar, in the presence of the Finance Minister, Shri. Yashwant Sinha and Dr. Manju Sharma, Secretaries, Department of Biotechnology. Biotech Consortium India Limited (BCIL) is a public limited company, promoted by the Department of Biotechnology (DBT),

Ministry of Science and Technology, Government of India and All India Financial Institutions for providing the necessary linkages among stakeholders and business support for facilitating accelerated commercialization of Biotechnology.

BCIL was incorporated under the Indian Companies Act 1956. The Board of Directors of BCIL consists of senior representatives of DBT, Council of Scientific and Industrial Research (CSIR), Indian Council of Agricultural Research (ICAR), leading all India financial institutions and the biotechnology industry.

BCIL has been actively involved in technology transfer, project consultancy, fund syndication, information dissemination, and worker training and placement related to biotechnology over the last decade and a half. It has assisted hundreds of clients, including scientists, technology research institutions, universities, the first entrepreneurs, the corporate sector, national and international organizations, central government, various state governments, banks and financial institutions.

On behalf of the Department of Biotechnology (DBT), Ministry of Science and Technology, Government of India, Biotech Consortium India Limited (BCIL) has been implementing since 1993–94, a scheme of practical industrial training. The objective of this program is to provide industry-specific training to students for skill development and to enhance their job opportunities in the biotech industry. This program provides an opportunity to the industry for training and selecting suitable workers. The constitution of BCIL is shown in Fig. 2.3.

Fig. 2.3 Constitution of BCIL

Linkages

By virtue of its unique constitution and various assignments it has taken up successfully for the national and international agencies, BCIL has close linkages with all the following major stakeholders responsible for developing the biotechnology industry in the country.

Confidentiality Agreements

Confidentiality clauses in technology transfer agreements serve to protect the proprietary nature of the technology. The drafting of the appropriate clauses can is essential for the maintenance of the value of the technology.

Establishment of a Confidentiality Agreement

The pre-contractual relationships which are directed toward the conclusion of a confidentiality agreement should identify the framework for negotiating the final contract. The draft confidentiality agreement should contain the necessary information about the parties and persons involved in the negotiations, a brief description of the confidential technology and of the means used for communications between the parties, Fig. 2.4. This draft should be circulated between the parties and complemented about any details, which may be helpful, in particular concerning the circle of persons involved on each side and the means for communications.

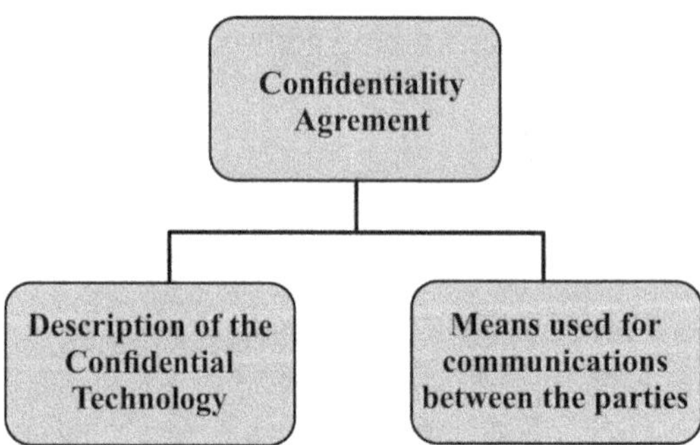

Fig. 2.4 The figure depicts the important information between the parties and the negotiating person

Description of Confidential Technology

During the pre-contractual phase, it is important not to reveal the essential elements of the technology through its description until a contractual partner has been found with whom the final contractual arrangement will be concluded. The description of the secret technology should disclose only as much information, which the recipient needs for the purpose envisaged. If negotiations are conducted with several potential recipients, it is essential to limit the description of the technology to avoid that those organizations with which no contract is concluded will not be able to exploit the technology.

Contractual Terms Concerning Confidentiality

The contractual terms, which bind a party to maintain a technology confidential is based on the principle of the freedom of contract. However, this freedom is, in general, limited in the public interest. Accordingly, obligations, which contravene the public order of a state are void, without effect and cannot be enforced before the courts of that state.

Confidentiality Clauses

The confidentiality clause may be a part of a contract. Such contracts may concern the acquisition of a plant or company, they may relate to a license for the use of *know-how* or a patented invention and other facts.

Applicable Law in International Contracts

If the parties to the agreement are established in different states, it may be recommendable if they choose the law applicable to the agreement. Since it can be difficult for a party to accept the application of the law of the state where the other party is established, they may consider the application of transnational rules of law or the settlement according to equitable principles.

If a contract and the subsequent exchange of information are made electronically, the parties should avoid doubts about the effects of the data exchange. It could be helpful if the parties agree upon the application of the UNCITRAL Model Law on Electronic Commerce, which defines, amongst others, the time of receipt of a data message and which establishes rules on evidence concerning the use of electronic data messages.

Penalty Clauses

Penalty clauses may be used to facilitate the obeisance with contractual obligations. Such clauses are particularly useful in the cases where it is difficult to calculate the amount of damages. However, considering into account the fact that at an early stage, the value of the Confidential technology will not be easily ascertainable, it may be difficult to arrive at an amount of damages for a breach of the contract or confidentiality, which is acceptable for both parties.

Licensing, Technology of Transfer

Technology transfer has been used in the movements of technology from the laboratory to industry or from one application to another domain application or taking developing countries into consideration technology transfer helps in growing access to technologies which are related to other developed countries and henceforth helps in approaching toward the newer technologies and inventions i.e. From developed to developing countries.

However, licensing is an allowance granted by the patent owner to another person or organization for using the patented invention on agreed terms and conditions, while the patent owner continues maintaining his ownership to the patent and hereafter becomes the source of income by receiving the predetermined royalties or as per the condition.

So by combining the concept of the technology transfer with the licensing one can help take the benefit of the technology research done previously, as licensing creates the permissible structure for the transfer of the technology to a larger assembly of researchers and engineers, which will help in saving the expenses of conducting the research and the costs of maintaining development activities or facilities and hence will help in the further development of the technology which has already been done.

As now a day in the era of the advancement in technologies, there are many technologies, which with the combination with the other technologies is giving birth to the other new advent technologies. So here the licensing does play an important role in providing the legal platform to use the combination of the technologies made or discovered by the other persons or the organizations created earlier, and hereafter prevents wastage of the time and the research costs incurred in developing the earlier inventions.

Form of Technology Transfer

Technology transfer can be classified into vertical and horizontal technology transfer.

Vertical technology transfer refers to the transfer of technology where the transmission of new technologies is done from the generation of new technology during the research and development programs into the science and technology organizations, for instance, to the application related to the industrial and agricultural sectors, or we can say that vertical transfer is the technology transfer starting from basic research to applied research, from applied research to development followed by development to production.

While the **horizontal technology transfer** is the movement of a well-known technology from one equipped environment to another (from one company to another) or say refers to the transfer and use of technology used in one place or organization to another place or organization.

As discussed above generally developed countries follow the route: -

Research −> Development −> Design−> Production

While less advanced and developing countries follow the route: -

Production −> Design −> Development −> Research

Generally, there are the reverse trends in the developing countries because the path to be followed depends upon the transfer, absorption, and adaptation of existing technology

Today in the era of the advent of technology, one could choose any route of the technology transfer, which depends upon how the technology advancement chains of the transferor and transferee are associated.

Indian Scenario Regarding the Licensing and Technology Transfer

Technology in India is growing exponentially and has played an important role in all round development and growth of the economy in the country, India has opted for a wise mix of original and imported technology. Henceforth "Technology transfer" plays an essential role and is generally covered by a technology transfer agreement.

Developing countries like India generally do not follow the usual path for development with regard to technology, but use their advantage in the cutting-edge technology options, which is now available and put the tools to use this modern technology.

Technology transfer has assumed to get benefits from R&D which is shared with both developing and underdeveloped countries, so taking this to the point of consideration National research laboratories is constructed by the Indian government for R&D, which are yet to be initiated by the private sector.

India generally comprises small and medium enterprises and has been grown since liberalization, which has resulted in the growth of the multinational enterprises, which in turn competes with the international companies, which has enhanced India's confidence. Not only confined to the pharmaceuticals but is broadly categorized in other areas too such as agriculture, dairy and other technologies.

Government of India is on the verge of opening Technology Transfer Offices, Universities, institutions which will be funded by the central government and will act as a mechanism for transferring or exporting the research conducted and its outcome to the desired place.

Though some of the Indian Institutes have already commercialized their research and are successful in technology transfer in which they have been licensed as technologies to industry. Moreover, numerous cases of technology transfer have been seen in India by various well-known institutions.

Technology transfer and its licensing have played a crucial role in all round development and the advent of the technology, which in results helps in the development of the economy of the country. Hence, forth helps in creating the wealth for the country.

India as a developing country must work on the technology development and technology transfer and must make a building strategy comprising the construction of new offices related to technology transfer and to make youngsters aware of the benefits related to the technology transfer, by establishing the specified universities and henceforth increasing the pace of the technology transfer and technical research and development in a technical perspective.

Legal Issues of Technology Transfer

Legal Contractual Agreements

The legal agreement contains details such as name, addresses and activities being done by both the parties, it clearly states the purpose of the agreement, scope, financial conditions, royalty rate, valid term of the agreements, details for arbitration in case of dispute, confidentiality term, provision for amendment in the agreement.

Legal Issues in Intellectual Property Transaction

Intellectual Property Rights for Patent, Trademark, copy right, trade secret, etc. Are rights parallel to those of real estate rights. Owner of IP can prevent the unauthorized use of his/her IP.

- "Making, using, offering for sale, selling, or importing without authorization" is called infringement. Two types of infringements namely Product infringement and process or method infringement are observed.

- IP can be characterized as right to prevent others from doing something. It need not be technology based but it should have commercial effects.

- The relief available against infringement includes are injunction, s damages or account of profits and seizure and destruction of infringed articles.

- Infringer must pay for the Damages, Lost profits, Reasonable royalty, Pre and Post judgment interest and Attorney fees.

- In cases of willful infringement, the court may increase the damages up to three times the amount found or assessed.

Problems Associated with IPR Litigation

- The court systems are ill equipped to handle the complex issues in infringement

- Judges and jury members have a lack of technical expertise. They are neither scientists nor engineers and typically spend between 0.01% to 2.0% of their total court time in patent litigation

- Lengthy trials and improper Judgments

- All litigation is expensive

Because of the inadequacies of resolving patent infringement disputes through litigation, a better/alternative method must be found and used.

The new method must rid itself of the time, costs, and quality deficiencies of litigation. Alternate Dispute Resolution (ADR) is "any mechanism for parties to resolve their dispute other than through traditional court litigation."

ADR is based on mutual contract/agreement. It is recognized by

- United Nations Commission on International Trade Law (UNCITRAL1980)

- Geneva Convention (1923/1961)

- The New York Convention (1958)

Legislations Covering IPRs in India

In India following IPRs are taken care of.

- Patents: The Patents Act, 1970 and was amended in 1999 and 2002.

- It was last amended in March 2005.

- Design: The Design Act 2000

- The Trade Mark: The Trade Marks Act, 1999

- Copyright: The Copyright Act, 1957 as amended in 1983, 1984 and 1992, 1994, 1999, and the Copyright Rules, 1958.

- Layout Design of Integrated Circuits: The Semiconductor Integrated Circuit Layout Design Act 2000.

- Protection of Undisclosed Information: No exclusive legislation exists, but the matter would be generally covered under the Contract Act, 1872.

- Geographical Indications: The Geographical Indication of Goods (Registration and Protection) Act 1999.

- Plant Varieties: The Protection of New Plant Variety and Farmers Rights Its Act 2001.

Patents, designs, trademarks and geographical indications are administered by the Controller General of Patents, Designs and Trademarks, which is under the control of the Department of Industrial Policy and Promotion, Ministry of Commerce and Industry. Copyright is under the charge of the Ministry of Human Resource Development. The Act on Layout Design of Integrated Circuits is administered by the Ministry of Communication and Information Technology. The Act on New Plant Variety is administered by Ministry of Agriculture.

A. Multiple Choice Questions

1. The technology acquisition means

    a) To gain for oneself through one's actions or efforts

    b) Value addition of technology

    c) Evaluation of technology

    d) Identification of technology

2. Which of the following is excluded from external technologies transfer?
    a) Licensing agreements   b) Enterprise acquisition
    c) Contracting agreements   d) Value addition

3. What do SU and RU stand for?
    a) Support unit and Receipt unit
    b) Sending unit and Receiving unit
    c) Support unit and Receiving unit
    d) Sorting unit and Receipt unit

4. The R&D department transfers the technology to the production department for
    a) Pilot-plant scale up
    b) Validation of the method
    c) Commercial production of the product.
    d) Development of analytical methods

5. Benchmarking helps understand
    a) Standardization of product quality
    b) Validation of the Master Manufacturing Formula
    c) Development of analytical standards
    d) How the drug product works comparatively with market samples.

6. Each product is prepared or manufactured through
    a) Single process
    b) Multiple processes
    c) Bracketing
    d) Change control

7. In technology transfer the process involved in post-approval phase occurs
    a) Recipe development   b) Route selection
    c) Site evaluation   d) Regulatory compliance

8. In technology transfer the process involved in pre-clinical phase is
    a) Sequence selection   b) Trail production
    c) Site evaluation   d) Regulatory compliance

9. Which of the following statements is correct?

    a) Primary packaging material must not be compatible with the drug product.

    b) Primary packaging material must prevent degradation of the medicine due to environmental influences.

    c) Primary packaging material must not release the content smoothly.

    d) Primary packaging material must not leach out any material.

10. Which of the following is a component of risk assessment?

    a) Identification of hazard   b) Analysis of the product

    c) Control of manufacture   d) IPQC

11. Which of the following is considered the technology transfer protocol?

    a) Key personnel and their responsibilities

    b) Identification of critical control points

    c) Experimental design and acceptance criteria for analytical methods

    d) All the above

12. Which is not the sources of uncertainty in risk evaluation?

    a) Gaps in knowledge,

    b) gaps in pharmaceutical science,

    c) Gaps in production

    d) process understanding

13. Which one of the following is not a non-exhaustive list of some of these tools?

    a) Failure Mode Effects Analysis (FMEA)

    b) Standard operating procedures (SOPs

    c) Fault Tree Analysis (FTA)

    d) Risk ranking and filtering

14. Which one of the following is not a process of technology transfer?

    a) Marketing protocol

    b) Quality Design (Research Phase)

    c) Scale-up and Detection of Quality Variability Factors (Development Phase)

    d) Technology Transfer from R&D to Production

15. The SU should provide a detailed characterization report of the product?

    a) Distribution details

    b) Marketing details

    c) In-process control method and specifications

    d) None of the above

16. The SU should not provide the information about

    a) Any health,

    b) Safety

    c) Environmental issues associated with the manufacturing processes to be transferred

    d) Manufacturing personnel

17. Which of the following statements is correct?

    a) Evaluation of quality information with the generation of site-specific stability data is not required.

    b) The transfer of packaging operations should follow the same procedural patterns as those of the production transfer.

    c) The transfer of packaging operations should not follow the same procedural patterns as those of the production transfer.

    d) All the above

18. Which one would be the main objective of the Intellectual Property Facilitation Centre?

    a) To provide value addition services regarding IP management and technology commercialization for making commercial products from the R&D of institutions in the region.

    b) Technology transfer is not classified into vertical or horizontal technology transfer.

    c) Horizontal technology transfer refers to the transfer of technology where the transmission of new technologies is done from the generation of new technology during the research and development programs into the science and technology organizations.

    d) Vertical technology transfer refers to the transfer and use of technology used in one place or organization to another place or organization.

19. In India, which of the patents right remains valid after the payment of required fees?

    a) 10 years
    b) 15 years
    c) 20 years
    d) 25 years

20. When was the Patent Act 1970 amended first?

    a) 1999
    b) 2000
    c) 1998
    d) 1997

B. Short Questions

1. What is technology transfer? What are the different types of technology transfer?

2. What is meant by internal technology transfer? What are the Barriers to Internal Technology Transfer?

3. Why is the External Technology Transfer done? What is meant by External Technology Transfer?

4. What are the general principles and requirements for successful transfer of technology in case of 'contract manufacturing'?

5. Write down the protocol to be followed for technology transfer.

6. What is general quality risk management? Write down the responsibilities of the team assigned for the said purpose.

7. What is Biotech Consortium India Limited and what are its function?

8. What do you about the confidentiality of the agreement?

9. Write a note on Intellectual Property Facilitation Centre.

10. Write down about legal issues in Intellectual Property Transactions.

C. Long Questions

1. What are the internal and external methods for technology transfer? What are the barriers to such a transfer and how can they be overcome?

2. What are the guidelines of WHO for technology transfer? How can the technology be transferred in case of 'contract manufacturing'?

3. Who approves the technology transfer process? Explain the processes involved in the technology transfer process.

4. Explain the principles of quality risk management. What is meant by WHO Guidelines on Quality Risk Management?

5. How can the quality risk be evaluated, controlled, and reduced?

6. Write down the information required in each specific case should be assessed using the principles of QRM with respect to the active pharmaceutical ingredient.

7. Briefly write down the information on the process and finished products required for safe and smooth transfer of technology.

8. Discuss briefly National Research Development Corporation and Technology Information, Forecasting and Assessment.

9. Explain the Indian scenario regarding the licensing and technology transfer.

10. Explain in brief the legal issues generally observed in technology transfer.

MCQs Answers

1	2	3	4	5	6	7	8	9	10	11	12	13	14	15	16	17	18	19	20
A	D	B	C	D	B	C	A	B	A	D	C	B	A	C	D	B	A	C	A

Regulatory Affairs & Regulatory Requirement for Drug Approval

- ➢ Introduction
- ➢ A Historical Overview of Regulatory Affairs
- ➢ Role of the Regulatory Affairs Department
- ➢ Responsibility of Regulatory Affairs Professionals
- ➢ Drug Development Teams
- ➢ Investigational New Drug (IND) Application
- ➢ Drug Approves Process in United States
- ➢ Categories and Areas of Investigational New Drug (IND)
- ➢ New Drug Application (NDA)
- ➢ Drug Approval Process in India
- ➢ Clinical Research and Bioequivalence Studies
- ➢ Objective of the Guideline
- ➢ Guideline for Bioequivalence Studies of Generic Products
- ➢ Biostatistics in Pharmaceutical Product Development
- ➢ Data Presentation for FDA Submission
- ➢ Management of Clinical Studies

Introduction

Drugs and pharmaceutical products have several unique attributes. These can distinguish drugs or pharmaceutical products from consumer products. Generally, the patients do not have the specialized knowledge required to make an independent assessment of the safety, quality and efficacy of the medicines. The information between the manufacturers, the doctors who prescribe drugs, and patients who ultimately consume these are not similar. Hence, need for regulatory supervision is widely acknowledged by all the stakeholders in the interest of public health. The quality and efficacy of the medicines also contribute to strengthening the faith in health systems, health professionals, pharmaceutical manufacturers and distributors in the company. The objective of all drug regulatory regimes is to ensure that safe, good quality and effective drugs that can reach the patients. Although the methods or mechanisms designed to fulfill these objectives can vary based on the nature and scale of the regulatory space that frames the operation of these regimes differ across countries. This makes various challenges to the principles of design and functioning of the regulatory structure. The regulation of drugs is a public policy to maintain the requirements of public health with the changing needs of pharmaceutical industry. The objective of regulatory control is a question of achieving a balance between protecting and promoting public health and facilitating the industry for compliance with regulatory standards. Although the regulatory objectives appear clear, the actual amount of regulatory oversight, the mechanism for achieving regulatory compliance and the actions required to perform non-compliance have to be designed in a manner that is sensitive to the characteristics of the regulatory space, and especially the actors in that space. The rationale behind this is to design a regulatory system where the choice of regulatory instruments not only match the imperatives/objectives of regulation, but also consider the range and the intrinsic characteristics of each regulatory stakeholder. In the Indian context, the architecture of drug regulation is designed as a classic commands and– control system in which the regulator prescribes standards, distributes licenses and then undertakes inspection to check for compliance. This includes various positive attributes such as clarity in regulatory standards. This makes it easier to apply and spot instances of non-compliance. However, such a system also requires considerable investment of resources, in areas ranging from setting standards to the maintenance of records, conducting inspections, collecting and testing samples, etc. To gain a global position in new drug discovery, India has started focusing on the manufacture and export of generic drugs. Lack of access to resources, both physical infrastructure and human resources, has continued to plague Indian drug regulators. This problem is more acute amongst the State Drug Regulatory Authorities. The Orissa State Drug Controller has requested the implementation of the Mashelkar Committee ratio in the Institutional Development Plan submitted to the Union Ministry of Health and Family Welfare (MOHFW). Therefore, the state of affairs has remained primarily unchanged for more than a decade, as the Mashelkar Committee report appeared. Ideally, the recommendations should have been implemented by now and indeed improved further. Because there have been many proposals to reform the department-related Parliamentary Standing Committee on Health and Family Welfare 59th report on the functioning of the Central Drugs Standard Control Organization (CDSCO) and recently, the Drugs and Cosmetics (Amendment) Bill, 2015.This is a first study of the legal architecture, administrative structure and functioning of drug regulatory authorities (CDSCO and SDRAs) in India, that focuses on;

1. The Functioning of Central Drugs Standard Control Organization (CDSCO), the national level regulator, and State Drug Regulatory Authorities (SDRAs) in India, which are governed by the Drugs & Cosmetics Act, 1940 (DCA);

2. Examining the nature and scale of the regulatory challenges facing the administrative structure and functioning of drug regulatory authorities in India.

3. Exploring the lessons that can be drawn from regulatory experience within the country and in other jurisdictions.

4. Evolving a set of actionable policy recommendations reflecting the views of range of stakeholders.

Therefore, the questions arose are the problems and challenges confronting the current drug regulatory system and potential mechanisms for addressing such problems and challenges. The CDSCO and the SDRAs are very closely tied to each other under the same ministries and departments of health, respectively. As a result, flexibility in decision-making and autonomy in a group of areas, including financial autonomy, recruitment and other areas of institutional policy is there. In fact, the most important requirement is the autonomy of the regulatory agency to greater flexibility and increase the operational effectiveness of both these regulatory agencies. There has been a lack of planning and execution of training programs for drug inspectors. A good number of SDRAs do not have a planned schedule for training programs

Since the nineteenth century, India has become mainly dependent on imported drugs. Due to the lack of availability of regulation in this country, there was a high quantity of adulterated, spurious and substandard drugs were in the market. To solve this problem, the Drugs and Cosmetic Act was enacted in 1940. This Act provides regulations for the import, manufacture, sale and distribution of drugs. The importance of the Drugs & Cosmetics Act is in the regulation of imported products. This explains the reason behind the lack of regulation of exports and lack of regulatory powers between the centre and the state/provincial governments. Hence, the licensing of imports was considered urgent and more important

Protection of public health is the fundamental purpose of Regulation. Regulatory affair is a profession, which protects the public health by controlling the safety and efficacy of products such as pharmaceuticals, medical devices, veterinary medicines, pesticides, agrochemicals, cosmetics and complementary medicines. The protection of public health is the primary goal and it can only be achieved through extensive and complex regulation. Safety, Efficacy, Risk/benefit and Quality are the core attributes on which regulation works.

Regulatory affair (RA) is a profession, which acts as the interface between pharmaceutical industry and drug regulatory authorities across the world. It mainly involves the registration of drug products in respective countries before their marketing. The current Pharmaceutical Industry is well organized, systematic, and compliant with international regulatory standards for manufacturing of Chemical and Biological drugs for human and veterinary consumption and medical devices, traditional herbal products and cosmetics. The department of Regulatory Affairs has become an important part of the organizational structure of pharmaceutical companies. Internally it links at the interface of drug development, manufacturing, marketing and clinical research. Regulatory affairs are actively involved in every stage of development of a new medicine and in the post marketing activities with authorized medicinal products.

Drug Regulatory Affair Department is the backbone of Pharmaceutical Industry. It is a revenue generator for Industry.

- The Regulatory Affairs department is an important part of the pharmaceutical companies. Internally, it coordinates with other departments like drug development, manufacturing, marketing and clinical research. Externally, it is the key interface between the company and the regulatory authorities.

- Regulatory affairs are involved in the development of new medicinal products, by applying regulatory principles and by preparing and submitting the relevant regulatory dossiers to health authorities.

- Regulatory Professional obtains the marketing authorization for the product by presenting the registration document to regulatory agencies, and making the necessary subsequent negotiations.

- They contribute to the commercial and scientific success of the product development by providing strategic and technical advice, from the beginning of the product development.

- More than 15 years span is required to develop and launch a new pharmaceutical product in the market. During this scientific development, many problems may arise.

- This process can be made fast by avoiding and solving the problem at appropriate steps only with the help of Regulatory Professional. They help the company in keeping proper records, appropriate scientific thinking. The departmental ways remain updated regularly with changing regulatory guidelines/requirements and makes proper presentation of data.

Legislations for marketing authorization and standards and quality of drugs are becoming more complex. Since earlier times, for specific market drugs are being developed keeping in view the regulations of that country or region. It is practically impossible to have drug products without fulfilling requirements of law. During initial marketing, it is impossible to get the marketing approval of the product without fulfilling the regulatory requirements. It is expected that post marketing surveillance data will be submitted and the variation application and renewal during the approval life cycle of the product.

However, it is becoming necessary to report the safety, efficacy, and quality of new medicinal products due to the rapid increase in laws, regulations, and guidelines. No manufacturing unit or marketing company can launch any drug in the market unless the respective health authorities (national/ international) give approval.

About two decades before, the pharmaceutical industry was little aware of drug regulatory affairs. It would have been very early stage to believe that registration executives can work under the export department. However, this situation has drastically changed. Full-fledged Global Regulatory Affairs department becomes mandatory to define drug development, approval, and marketing strategy. Hence, the scope of drug regulatory affairs has become wide, and the experts are required in health authorities and pharmaceutical industry at various levels and departments. A union is formed by amalgamating various regions so that the drugs of the same quality become available due to prevailing of a harmonized regulatory requirement. The harmonized unions are the European Union, Gulf Co-operative Countries (GCC), Association of Southeast Asian Nations (ASEAN). In most of the markets such as regulated, semi-regulated or less regulated, the technical requirements are the same, that is 18 – 24 months; thus, it becomes difficult to segregate these markets. Similarly, the regulation pertaining to pharmacovigilance requirements mandate companies to monitor new drugs safety aspects throughout the lifecycle of the product. Under such circumstances, the role of regulatory experts becomes very critical and important in deciding the entry strategy into various national and international markets.

Therefore, the drug regulatory affairs can be defined as *a function that regulates the pharmaceutical science to facilitate trade/business in and outside the country of origin for the public interest.* However, drug regulatory affairs regulate the pharmaceutical business through designing appropriate laws (rules) and enforcing the same so that the drugs of the highest quality are brought to the global trade market. Rules and regulations are accordingly being prepared considering the global, regional, and national pharmaceutical markets and the necessity of the drugs based on a patient population. Most of the national guidelines for developing drug and marketing authorization application have been defined based on global and regional harmonized guidelines. Rules and regulations are being prepared based on global, regional, and national pharmaceutical trade, and on patient population. For drug development and marketing authorization application, most of the national guidelines have been defined as per global and regional harmonized guidelines. Subsequently, in collaboration with USA, EU and Japan, ICH issued Harmonized technical requirements for manufacturers to follow for MAA in these three regions. The World Health Organization (WHO) comprising the countries other than the USA, EU and Japan, and prepared harmonized MAA guidelines and technical data requirements for registration in all countries.

A Historical Overview of Regulatory Affairs

Drugs and ordinary consumers' products are not the same. In most cases, consumers cannot make any decisions about when and how to use drugs. The therapeutic benefits against risks must be known since there is no medicine, which is completely safe. Hence, necessary professional advice from either prescribers or dispensers is required to making these decisions. During the 1950s, many tragedies happened due to the misjudgement of the personnel during manufacture and some intentional addition of adulteration of substances into the pharmaceutical products. This has led to the deaths of these patients. After so many incidents, the regulatory bodies introduced the new laws and guidelines to improve the quality, safety, and efficacy of the products. This has also resulted in stricter norms for Marketing Authorization (MA) and Good Manufacturing Practices (GMPs). In the 1970s, the regulatory profession began to emerge as a health-related profession. Then in the 1980's it recognized internationally and the demand for it grew. In 1991, Regulatory Affair Certification (RAC) was introduced.

The report of the Mashelkar Committee was published in 2003. It remains one of the most important reform proposals with respect to the regulation of drugs in India. An immediate critical report on the growing circulation of spurious and adulterous drugs in India. For offences related to spurious drugs, including making them cognisable and non-bailable offences, the Committee made specific recommendations. Penalties to be made such as more fines and imprisonment for life depending on the gravity of the crime. Special courts are currently being established for fast-tracking such cases. Moreover, it highlighted the challenges that the drug regulatory system faces. These included the insufficiency of trained and skilled personnel at the centre and stake levels, lack of uniformity in the implementation of regulatory requirements and variations in regulatory enforcement, the lack of a database of drug products licensed and inadequacy and lack of coordination among drug testing laboratories. The reports of the Mashelkar Committee made several recommendations for examining the system.

- To renew the CDSCO to make it strong, well-equipped, empowered, independent and professionally managed, it could be given the status of Central Drug Administration (CDA) reporting directly to the Ministry of Health.

- It categorized the states in terms of the scale of industry (manufacturing and salle) and regulatory operations and pleaded for a differential approach in terms of public investment in regulatory resources.

- It suggested the revision and imposition of higher fees for drug applications, clinical trials and registration of imported drugs and foreign manufacturers.

- It worked out a formula for adjudging the required personnel. For effective implementation of the laws, it recommended that there should be one drug inspector per 50 manufacturing unit and per 200 sales/distribution outlet.

Interestingly, the central government and to a certain extent state government implemented all these recommendations with reference to spurious drugs. However, the other recommendations have been largely ignored. After the Mashelkar Committee report in 2013, the 59th Parliamentary Committee report was published. It made a formal accusation of crime in the functioning of CSDCO and highlighted the regulatory failures brought on by the non-adoption of the recommendations of the Mashelkar Committee. The main recommendations of the 59th Parliamentary Committee report include.

1. Reformulating the mission statement of CDSCO, which has been implemented.

2. Hiring short-term consultants with clear conflict of interest guidelines to address recurrent staff shortages due to staggered recruitment through UPSC.

3. Priotising clinical trials over new drug licenses in terms of investment of regulatory resources given that the scale of the former is much larger than the latter.

4. Review the qualification, the procedure of selection, appointment, tenure, emoluments, allowances, and powers, both administrative and financial, of the DCGI.

5. Following the Mashelkar Committee formula for appointing adequate regulatory personnel.

6. Deciding the financial package for State Drug Authorities to upgrade infrastructure facilities, especially drug laboratories and coordinating between SDRAs through a centralized database.

7. Conducting Phase IV trials of approved drugs to locate side effects instead of relying on spontaneous reporting.

8. Widely publicizing the news of substandard drugs through press releases and paid newspaper advertisements.

In response to this, the MOHFW constituted an expert committee under Professor Ranjit Roy Chaudhary (RRCC) to develop policies and guidelines for the approval of drugs, clinical trials and banning of drugs. The RRCC has also made recommendations for developing standard operating procedures (SOPs) and establishment of technical expert committees for new drug approvals, registration of institutional ethics committees, firm timelines for processing of applications and clarification on the public interest waiver for clinical trials and weeding out of hazardous and irrational drugs.

The current Pharmaceutical Industry is well organized, systematic, and compliant with international regulatory standards for manufacturing of Chemical and Biological drugs for human and veterinary consumption and medical devices, traditional herbal products and cosmetics. Stringent GMPs are being followed for blood and its derivatives as well as controlled manufacturing for Traditional Herbal Medicines, Cosmetics, Food and Dietary products, which was otherwise differently a century before. Each regulatory system had faced certain circumstances, which led to the current well-defined controlled regulatory framework. This has resulted in the systematic manufacturing and marketing of safe, effective and qualitative drugs. With the growth of industry, the legislation from each region has become more and more complex and created a need for regulatory professionals. To understand the chronological development of the modern era of pharmaceutical industry and the regulatory framework, we will glance through the historical evolution of regulations in the USA, Europe and India. Objectives After reading this paper, you should be able to understand - How and why the pharmaceutical industry and drug regulations have developed in USA - Major Regulations of USA - EU and its regulatory framework - Content of "The Rules Governing Medicinal Products in the European Union" - Pharmaceutical Legislations of EU - EMEA and its Committees - Types of Marketing Authorization Procedure in EU - Indian Pharmaceutical Industry & Drug Regulations development in different Era - Major Rules and Act of India - Drug Regulatory Affairs and Global, Regional and National Regulatory Network - Roles of Regulatory Affairs Professional in Health Authorities as well as Pharmaceutical Industry - Regulatory Affairs Network in Pharmaceutical Industry Historical Overview of Pharmaceutical Industry During 1950s, multiple tragedies i.e. Sulphanilamide elixir, vaccine tragedy and thalidomide tragedy has resulted in a substantial increase in legislation for drug product quality, safety and efficacy. This has also resulted in stricter norms for Marketing Authorization (MA) and Good Manufacturing Practices (GMPs). Let us see what happened in the USA, Europe, and India.

Regulation means "controlling human or societal behaviour by rules or regulations or a rule or rule/order issued by an executive authority or regulatory agency of a government and possessing the strength and force of law."

Collecting, analyzing, and communicating the risks and benefits of health care products to regulatory agencies and the public all over the world. It is also a science of developing new tools, standards, and approaches to evaluate the safety, efficacy, quality, and performance of regulated products. All medicines must meet three criteria:

➤ Good quality,

➤ Safety, and

➤ Effectiveness

The decision on quality, safety, and effectiveness should be based on solid science. The success of regulatory strategies is less dependent on the regulations than on how they are interpreted, applied, and communicated within companies and to outside constituents. Professionals in pharmaceutical regulatory affairs play a primary role in ensuring compliance with regulations governing these industries. Those who work in pharmaceutical regulatory affairs jobs work in the initial application phase for a new or generic drug, in the licensing and marketing stages – ensure sure all operations and products meet the required safety and efficacy standards. By combining knowledge about the business, legal, and pharmaceutical industries the professionals can determine whether the regulations are being followed and, in most cases, form the link between pharmaceutical companies and regulatory authorities, such as the Foods and Drugs Administration (FDA) and the European Union.

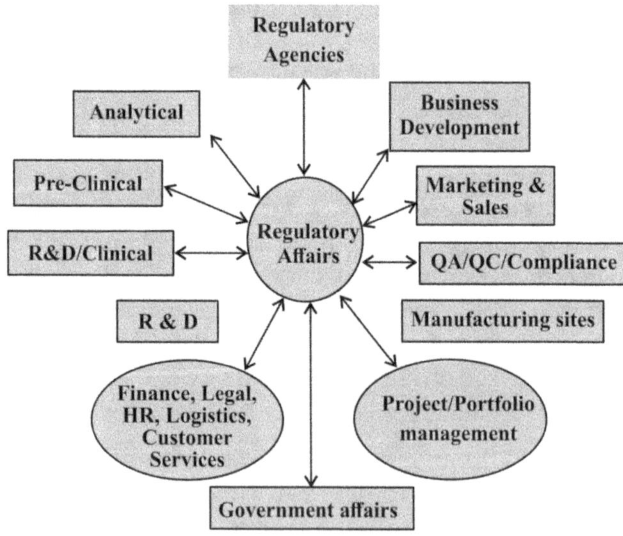

Fig. 3.1 Roles of regulatory affairs

Some Important Drug Regulatory Authorities (Across the World)

Each and every country has its own regulatory body. Some important authorities among them are mentioned below along with their country name.

- CDSCO- India
- USFDA- United States of America
- EMA-European countries
- MHRA- United Kingdom
- Health Canada- Canada
- TGA- Australia
- MCC- South Africa
- ANVISA- Brazil
- MEDSAFE- New Zealand

Central Drugs Standard Control Organization (CDSCO)

CDSCO is the Central Drug Authority for discharging the functions assigned by the Ministry of Health & Family Welfare, Directorate General Health Services, Government of India under the Drugs & Cosmetics Act. CDSCO has six zonal offices, four sub-zonal offices, 13 port offices and 7 laboratories under its control.

Thus, as per the Drugs & Cosmetics Act, 1940 and rules 1945, both central and state authorities have been entrusted to enforce the regulation of drugs and cosmetics. According to the law uniformity, the implementation of the provisions of the Act and Rules there under ensures the safety, rights and wellbeing of the patients by regulating the drugs and cosmetics. The CDSCO has been constantly trying to bring out the transparency, accountability and uniformity in its services to ensure the safety, efficacy and quality of the pharmaceutical products manufactured, imported and distributed in the country. Under the Drugs and Cosmetics Act, the CDSCO is responsible for the approval of drugs, the conduct of clinical trials, laying down the standards for drugs. To control the quality of imported drugs in the country and to coordinate the activities of State Drug Control Organizations provide expert advice to bring about the uniformity in the enforcement of the Drugs and Cosmetics Act.

The Drug Controller General remains responsible primarily for

- Blood,
- Blood products,
- I.V. fluids,

- Vaccines and
- Sera.

However, the CDSCO along with state regulators stand jointly responsible for granting of licenses of the above-mentioned specialized categories of critical drugs.

According to the Drugs and Cosmetics Act, the State Authorities are routinely responsible for

- Manufacture,
- Sale and
- The distribution of drugs in the state.
  The Central Authorities are responsible for
- Approval of New Drugs,
- Clinical Trials in the country,
- Laying down the standards for drugs,
- Controlling the quality of imported drugs,
- Coordinating the activities of State Drug Control Organizations and
- We are providing expert advice to bring about the uniformity in the enforcement of the Drugs and Cosmetics Act.

The functions of the CDSCO have been depicted in Fig. 3.2 below.

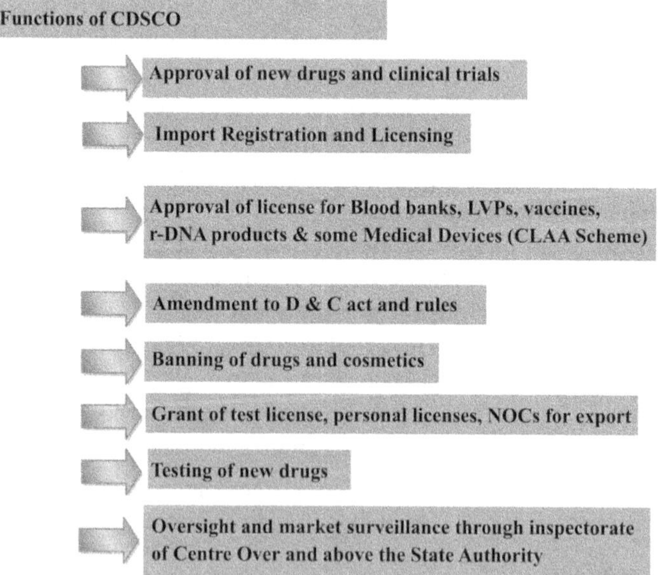

Fig. 3.2 Functions of CDSCO

Scope of Regulatory Affair

All the pharmaceutical industries are supposed to obey the rules and regulations framed by the regulatory authority. Thus, Regulatory Affair has great scope anywhere in the world, including India. Some important scopes are mentioned below.

- The proper development of the Product.

- Assuring efficacy, safety, and quality of the product.

- Review in-house documentation and updating of other departments.

- Proper packaging and labeling.

- Dossier filling/product submission and obtaining approval from authorities within a specified period.

- Making smooth roadways for placing quality products in the market.

- Helping Pharmacovigilance of quality/safety issue post marketing.

Role of the Regulatory Affairs Department

As a discipline, the regulatory affair envelops several specific skills and occupations. Under the prevailing circumstances, it is composed of a group of people who work as a liaison between the potentially conflicting worlds of government, industry, and consumers to ensure that the marketed products are safe and effective when used as claimed. People representing the regulatory affairs negotiate the interaction between the regulators (the government), the regulated body (industry), and the market (consumers) to get good products to the market and to keep the good products for sale and the bad ones are recalled. The range of products covered may be wide, such as foods and agricultural products, veterinary products, surgical products, and medical devices, in vitro and in vivo diagnostic tools and tests, and drugs including small molecules of proteins. The range of activities is wide such as manufacturing and testing, preliminary safety and efficacy testing, clinical trials, and post marketing follow-up. Additionally, issues related to advertisement, data management of healthy dose, document preparation, project management budgeting, issues related to negotiation, and conflict resolution, etc.

The regulatory affairs (RA) department of a pharmaceutical company remains responsible for obtaining approval for new pharmaceutical products. It is to ensure that approval is maintained for as long as the company wishes to keep the product in the market. The department and its officers serve as the link between the respective regulatory authority and the project team. It is the channel of communication with the regulatory authority as the project proceeds. The department aims to ensure that the project plan goes on correctly as required for the approval of the product.

It is the responsibility of the RA to remain aware of the current legislation, frame the guidelines for other regulatory intelligence. Such guidelines often provide some flexibility. It becomes the responsibility of the management of a company to interpret their decisions. The RA department plays an important role in providing advice to the project team on how best they can interpret the rules. During development phase sound working relations with authorities are required; such as to discuss such issues as divergence from guidelines, the clinical study programs, and formulation development.

Most of the companies assess and prioritize the new projects on the basis of their Target Product Profile (TPP). The RA professional plays an important role in advising on what will be practical prescribing information (label) for the intended product. As a member of the project team, RA also helped in the design of the development program. The RA department reviews all the documents received from the regulatory perspective. This is done to ensure that the information provided is clear, consistent and the document is complete in all respects including explicit conclusions. The department also prepares the draft containing the prescribed information on the basis of global requirement for approval so that the product could be marketed later on. The documentation includes the applications for clinical trials, as well as regulatory submissions for new products and for changes to the product already approved. The latter is a major task and accounts for about half of the work of the RA department. An important but proactive responsibility of the RA is to provide input when legislative changes are being discussed and proposed. In the ICH environment, there is a greater possibility of exerting influence at the preliminary stage.

On the government side, the people in regulatory affairs work to interpret and implement laws so that the interest of the users or consumers of drugs can be established. The Food and Drug Administration (FDA) requires the pharmaceutical companies to generate and provide all information supposed be necessary to evaluate a given drug, biological, and/or device with respect to safety and efficacy. Based on such information, the local drug regulatory authority decides whether the product will be on the market or not. The fig 1 shows the functions of the regulatory agencies.

Responsibility of Regulatory Affairs Professionals

- To keep in touch with international legislation, guidelines and customer practices

- To remain updated with a company's product range.

- To ensure that a company's products comply with the current regulations.

- To keep track of the ever-changing legislation in the regions in which the company wishes to distribute its products. They also advise on the legal and scientific restraints and requirements, and collect and evaluate the scientific data generated in their research and development department.

- To develop regulatory strategy for all appropriate regulatory submissions for domestic, international and/or contract projects.

- To coordinate, prepare and review all appropriate documents, for example dossier, and submit them to regulatory authorities within a specified time frame in consultation with the organization.

- To prepare and review SOPs related to RA. A review of BMR, MFR, change control and other relevant documents.

- To monitor the progress of all registration submission.

- To maintain approved applications and the record of registration fees paid against the submission of DMF's and other documents.

- To respond to queries as they arise, and to ensure that registration/approval is granted without delay.

- To impart training to R&D, Pilot plant, ADI, and RA team members on current regulatory requirements.

- To advise the company on the regulatory aspects and climate that would affect the proposed activities; that is, describing the "regulatory climate" around issues such as the promotion of prescription drugs and Sarbanes-Oxley compliance.

- To manage review-audit reports and compliance, regulatory and customer inspections.

- To help the company avoid problems caused by poorly maintained records, inappropriate scientific thinking or poor presentation of data. In most product areas where regulatory requirements are imposed, restrictions are also placed upon the claims which can be made for the product on labelling or in advertising.

- To provide physicians and other healthcare professionals with accurate and complete information about the quality, safety and effectiveness of the product.

Drug Development Teams

The drug discovery and development process requires the close interaction of a large number of scientific disciplines for many years. Most pharmaceutical and biotechnology firms appoint teams to guide the processes involved in taking a discovery through the various drug development stages and making the drug candidate into a therapeutic product.

The primary responsibility of this group is to identify lead compounds or classes of compounds worthy of continued research. They include other scientific disciplines such as synthetic chemists, therapeutic disease groups (cardiovascular, CNS, cancer. infectious diseases, or metabolic pharmacologists or biologists). Once a lead or class of leads has been identified, the discovery team further includes bioanalytical chemists, pharmacokinetics, and toxicology experts for the development of the drug candidate.

The responsibility of this group is to establish the preclinical development of a lead candidate in coordination with other new teams. These include, a project team leader assigned by the management, and coordinator, RA and QA professionals, clinical trial team, analytical chemists, manufacturing and marketing analysts. One of the final responsibilities of the preclinical drug development project team is to prepare an IND and to submit the appropriate documents to the FDA or other regulatory agencies for their information and necessary action. As the development of the drug candidate moves into the clinic, the preclinical project team becomes a clinical project team.

After the IND is submitted, the project team is again expanded to include physicians, clinical research associates, drug product production chemists, QA representatives, statisticians, and representatives from clinical pharmacokinetics, regulatory team medical monitors, marketing departments.

Manufacturing team, concurrently work with various other teams, during nonclinical and clinical drug development stage which includes pilot plant and scale-up process, validation, regulatory documentation, packaging, and labelling requirement according to GLP or GMP guidelines.

This team determines whether a drug candidate has potential in comparison to others company's product or drugs already in the market for indicating the disease. Further, Phase IV studies are undertaken after marketing has begun and evaluated new indications, new doses or formulations, long-term safety, or cost-effectiveness. These include the identification and monitoring of new additional adverse drug (ADR) events from doctors or other health professionals

The overall activities of a Drug Development Team can be presented as

1. Reviewing the results of research or experiments conducted by different scientific disciplines.

2. Integrating the outcomes of research from previously generated data and new research.

3. Planning the research studies for characterization of a drug candidate

4. Preparing a detailed plan for drug development, including the designation of key points or development milestones, generating a timeline for completion, and defining the critical path.

5. Monitoring the research activities to ensure that they are being conducted according to the timeline and critical path in the development plan and modifying the plan as new information becomes available

6. Comparing research outcomes, developmental status, and timelines with drug candidates which are under development stage with the competitors available.

7. Conducting appropriate market surveys to ensure that the development of a drug candidate is economically justified and continues to meet a medical requirement.

8. Reporting the status of development of drug program to the management and making recommendations on the continued development of the drug candidate.

Documents Prepared by the Regulatory Professional

The drug regulatory professional prepared the following documents:

- Internal Regulatory Standard Operating Procedure (SOP)

- New Drug Application (NDA), Abbreviated New Drug Application (ANDA),

- Investigational New Drug Application (INDA) - submission of documents.

- Drug Master File (4US type I to V, Canadian, Open, and closed part for EU and other market)

- Dossiers for submissions- 5 Common Technical Documents (CTD)/Asian CTD (ACTD)/non-eCTD electronic submission (NeeS)/6Electronic CTD (eCTD)/Country-specific dossier.

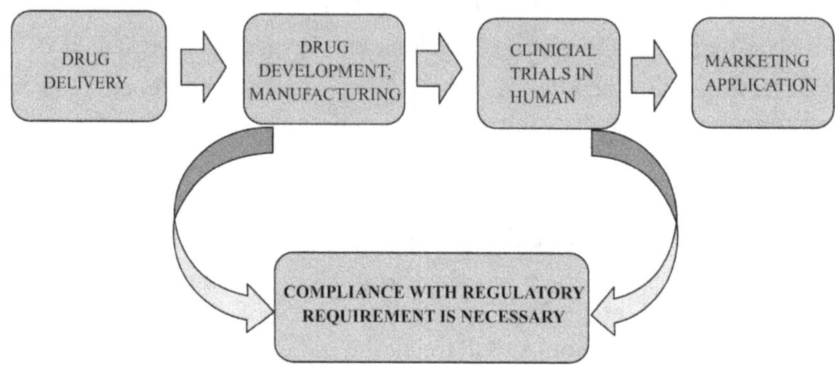

Fig. 3.3 Regulation of drug approval process

Currently, different countries must follow different regulatory requirements for approval of new drug. For marketing authorization application (MAA) a single regulatory approach applied to various countries is almost a difficult task. Therefore, it is necessary to have knowledge about the regulatory requirements for MAAs in each country. The basic regulation can be understood from figure 1given.

New Drug Application (NDA)

New drug application (NDA) is an application submitted to the respective regulatory authority to grant permission to market the drug. To obtain this permission a sponsor submits preclinical and clinical test data for analysing the drug information and description of manufacturing procedures.

Different phases of clinical trials:

- Pre-clinical study - Mice, Rat, Rabbit, Monkeys.
- Phase I - Human pharmacology trial - estimation of safety and tolerability.
- Phase II - Exploratory trial - estimation of effectiveness and short-term side effects.
- Phase III - Confirmatory trial - Confirmation of therapeutic benefits.
- Phase IV - Post marketing trial - Studies done after drug approval.

After NDA received by the agency, it undergoes a technical screening. This evaluation ensures that sufficient data and information have been submitted in each area to justify "filing" the application. At the conclusion of the review of a NDA, there are 3 possible actions that can be sent to the sponsor:

- Not approvable – in this letter list of deficiencies and the reasons there of are explained.

- Approvable – it means that the drug can be approved with minor deficiencies that can be corrected like-labelling changes and possible request commitment to conduct post-approval studies.

Approval – it states that the drug is approved. If the action taken is either an approvable or a not appropriate, then the regulatory body provides the applicant with an opportunity to meet the agency and discuss the deficiencies.

A new drug application (NDA) can be filed only when the drug successfully passes all three phases of clinical trials, including all animal and human data, data analyses, pharmacokinetics of drug and its manufacturing and proposed labelling are satisfactory. The preclinical, clinical reports and risk-benefit analysis (product's beneficial effects outweigh its possible harmful effects) are reviewed at the Centre for Drug Evaluation and Research by a team of scientists. If clinical studies confirm that a new drug is relatively safe and effective, and will not pose unreasonable risks to patients, the manufacturer files a New Drug Application (NDA), the actual request to manufacture and sell the drug in the United States.

Generally, the approval of an NDA is granted within two years (on an average); however, this process can be completed from two months to several years. The innovating company is allowed to market the drug after the approval of an NDA and is considered as in Phase IV trials. In this phase, new areas, uses, or new populations, long-term effects, and how participants respond to different dosages are explored.

Early Studies in Drug Discovery

The drug discovery industry typically follows one of two models to identify hits and lead compounds: the phenotype (or physiology) or the target-based approach, which differ in the way that they lead to therapeutic target identification and small compounds selection/optimization. The therapeutic target is a broad term that indicates the site where the substance will bind/act promoting its biological activity. In the phenotype screening approach, "forward pharmacology" or "classical pharmacology," substances are characterized according to their physiological effects based on disease-relevant assays (e.g., cell-based phenotypic assay, isolated tissue or animal models of disease) developed using basic biological and disease knowledge or clinically effective drugs. This approach can potentially lead to the identification of a molecule that modifies the disease phenotype through interaction or modulation of a previously undefined target (receptors, transcription factors, cytoskeleton proteins, enzymes and others) or simultaneously on multiple targets. In the target-based approach (also known as "reverse pharmacology" or "reverse chemical biology"), the goal is to develop drugs that affect the specific and known target, which are mostly proteins (receptors, enzymes, transporters, ionic channels, intracellular receptors, etc.) Previously discovered and linked to the human disease. The early steps of the target-based approach consist of target identification and validation. The drug design process occurs frequently, but not necessarily, by using computational modeling approaches that can selectively interact with the identified target.

Identification and Validation of Therapeutic Targets

The targets on target-based drug discovery are identified using several molecular tools and strategies that include the evaluation of DNA/RNA (genomic) and proteins (proteomic) that have been correlated with a human disease. On the phenotypic-based approach, the activity of substances is previously observed and then the target is identified. For this reason, efforts have been made to clarify the target or targets responsible for the phenotypic effects after obtaining the lead substances. The process of target identification in phenotype-based approach is also known as deconvolution. Target deconvolution can be obtained by chemical proteomic-based approaches (affinity chromatography, activity-based protein profiling, label-free techniques), expression cloning techniques, *in silico* approach and others. After identification, the therapeutic target should be validated. The aim of the validation is to evaluate whether the modulation of the therapeutic target is able to generate a reasonable biological response. Validation techniques range from in vitro tools to the use of whole animal models, and to modulation of a desired target in diseased patients. Nevertheless, target validation is not a one-step experiment, but an on-going part of a strategy program, which begins with target identification and is not complete until the definitive clinical study. The most accepted criteria for target validation during drug discovery are based on three categories:

Demonstration of the target protein expression or relevant cell types or in the target tissues from animal models or patients,

In-silico Assays

The term '*in silico*' is used to define experimentation performed by computers and is related to the more commonly known biological terms in vivo and in vitro. It defines the use of information in the creation of computational models or simulations that can be used to make predictions, suggest hypotheses, and ultimately provide discoveries or advances in medicine and therapeutics. The advantages of in silico studies are the speed of execution, the low cost and the ability to reduce the use of animals. The use of *in silico* methods has been an interesting strategy to accelerate the discovery of potential new drugs. The design of *in silico* drug prototypes ranges from the study of the structure–activity relationship to toxicology and pharmacokinetic studies (ADME: absorption, distribution, metabolism, and excretion).

High-throughput Screening Assay

For about two decades, the identification process of new substances with activity on a specific target has been slow, difficult and with limited yield. The advent of genomic sciences, rapid DNA sequencing, combinatorial chemistry, cell-based assays and automated high-throughput screening (HTS) has led to a "new" concept of drug discovery.

Currently, HTS is a well-established process in lead discovery for pharma and biotech companies and is now being set up also for basic and applied research in academia and hospitals. HTS operations are highly automated and computerized to handle sample preparation and assay procedures, so a large number of hypothetical targets can be incorporated into molecular or cell-based assays and exposed to many compounds representing numerous variations in a few chemical classes, which can certainly accelerate the generation of lead substances.

Proof-of-concept (or Principle) and Study of the Mechanism of Action of a Substance

The development of a new drug requires a meticulous and strategic scheme of different evaluations that encompass a chain of events that may overlap and transcend the non-clinical and clinical investigation steps. The evaluation of non-clinical activity (efficacy) of a new drug candidate includes *in vitro, ex vivo* and *in vivo* assays that can be conducted out throughout all stages of drug development. These tests are essential to provide the basic knowledge about the drug pharmacodynamics and are required for clinical studies. Due to improvements in the basic sciences of the biomedical field, it is natural that the methods for analysing the non-clinical efficacy of a drug candidate are thus constantly advancing. Although it is currently essential to use experimental animals, efforts have been made to reduce the use of in vivo techniques to a minimum.

The ARRIVE (from Animal Research: Reporting of In Vivo Experiments) guideline, which describes several important steps in the planning and execution of non-clinical efficacy tests, such as proper experimental design, powerful analysis techniques and thorough evaluation of the results. Despite efforts for the standardization and validation of alternative methods, in vitro assays for evaluating non-clinical efficacy have major limitations and the use of experimental animals in the process of drug development is still essential to meet the standards required by the main agencies that control the registration and use of drugs.

In vitro and *Ex-vivo* Assays

Usually, in vitro assays are performed in the early stages of the drug discovery process, when the selectivity and possible interactions of the candidate substance towards the desired therapeutic target are established. At this step of drug development, several molecules, previously selected by *in silico* methods and/or by HTS screening, can be tested. After the selection of promising substances, one may conduct further in vitro tests for the initial evaluation of biological activity. However, it is important to state that in vitro evaluations carried out in cells or in an individual target, such as an enzyme, can eventually provide false results. For this reason, the candidate drug should also be evaluated by ex vivo models, in which more complex structures are considered. In these models, some animals are used for biological material harvest, for example, smooth muscle of the gastrointestinal tract, airways, urinary tract, blood vessels, brain, cardiac muscle, endocrine glands, liver, and spleen, among others.

In vivo Assays

Based on the initial results of in vitro assessments and ex vivo studies, as well as on information about the therapeutic target, the clinical indications, and the knowledge of the pharmacokinetic profile of the candidate substance, it is generally outlined the in vivo tests are with the aim of determining the efficacy in a particular biological model. Furthermore, the selection of which in vivo experiments will be conducted should be discussed on a case-to-case basis and the most powerful and scientifically relevant methods should be selected. Generally, disease models can be divided into three types:

1. Those which are induced by an invasive procedure (physiological),

2. Those which are induced by a substance (pharmacological), and

3. Those using genetically modified animals (genetic).

All these models are designed to develop an abnormality similar to which occurs in the disease under investigation. Moreover, the in vivo models can be subdivided into acute and chronic, depending on the duration of the disease.

Reproducibility, Quality and Reliability of Non-clinical Studies

To achieve high quality in non-clinical studies, it is imperative that the institution works in accordance with the institutional Good Scientific Practice, which is analogous to the Good Laboratory Practice (GLP). The GLP principles are a quality program related to organizational processes and experimental conditions, required for a non-clinical study to be accepted in the main international regulatory agencies. Some critical points should be considered to eliminate any potential deviations that may interfere with and/or produce false results and affect the reliability, quality and reproducibility of the results of non-clinical studies, such as

1. Sanitary quality of experimental animals;

2. Animal distribution in the different experimental groups;

3. Sample size calculation;

4. Management of experiments;

5. Appropriate statistical analysis;

6. Correct data handling;

7. Quality of the reagents; and

8. Need for negative and positive control groups

Early studies for the identification and validation of therapeutic targets and the *in silico* and high throughput screening approaches are briefly discussed, as well as the proof of concept (or principle) assays that contributes to lead drug candidate selection, which will advance the development process.

Non-Clinical Studies in the Process of New Drug Development

The process of drug development involves non-clinical and clinical studies. Non-clinical studies are conducted using different protocols including animal studies, which mostly follow the Good Laboratory Practice (GLP) regulations. A drug candidate needs to pass through several steps, such as determination of drug availability (studies on pharmacokinetics), absorption, distribution, metabolism and elimination (ADME) and preliminary studies that investigate the candidate safety, including genotoxicity, mutagenicity, safety pharmacology and general toxicology. These preliminary studies generally do not need to comply with GLP regulations. These studies aim at investigating the drug safety to obtain the first information about its tolerability in different systems relevant for further decisions. There are, however, other studies that should be performed according to GLP standards and are mandatory for the safe exposure to humans, such as repeated dose toxicity, genotoxicity and safety pharmacology. These studies must be conducted before the Investigational New Drug (IND) application.

These observations and deficiencies became public and, Introduction to Good Laboratory Practice

The formal concept of "Good Laboratory Practice" (GLP) was launched in the USA, during the 1970s, owing to constant discussions on the robustness of the nonclinical safety data submitted to the FDA for New Drug Applications (NDA). At that time, inspections performed in the laboratories revealed

Inadequate planning and flaws in studies execution;

- Insufficient documentation of methods, results and even fraudulent data (for example, the replacement of dead animals during a study by other animals not properly treated by the test article), which were not documented;

- Use of hematological data from other studies as a control group;

- Exclusion of data concerning macroscopic observations (necropsy);

Raw data changed to "adjust the results with the political effect of these claims, the FDA published the first proposals for regulation in 1976 and the final rules in 1979. The GLP principles were the basis for ensuring that the reports submitted to the FDA fully and reliably reflected the experimental work conducted.

For the registration of pesticides, the US Environmental Protection Agency (EPA) found similar problems with the quality of the studies. Thus, the EPA published its own regulation draft in 1979 and 1980, and later, in 1983, the final rules were published in two separate parts; insecticides, fungicides and rodenticides and control of toxic substances, reflecting their different legal bases.

In Europe, the OECD (Organization for Economic Co-operation and Development) established a group of experts to develop the first GLP principles. This was an attempt to:

- Avoid non-tariff barriers to the marketing of chemicals;

- Promote mutual acceptance (among member countries) of non-clinical safety data;

- We eliminate the unnecessary duplication of experiments.

Concepts for Good Laboratory Practice

The concept given the term GLP can be considered an example of concise and precise definition, which not only defines GLP as a quality program, but differentiates it from other systems. GLP restricts actions to organizational processes, where all steps related to non-regulated clinical trials should be developed, from conception and design to the final stages (preparation of the final report and archiving). Thus, GLP is defined as *"a quality program related to organizational processes and conditions where non-clinical health and environmental safety are planned, performed, monitored, recorded, reported and archived."*

GLP is based on three main ways:

(i) **Test Facility Management (TFM): the person(s) with the authority and formal responsibility for the organization and good functioning of the operational unit in accordance with the principles of GLP.**

(ii) **Quality Assurance Unit (QAU): the QAU is an internal system designed to ensure that the GLP principles are met and that the studies conducted in the installation test are in accordance with these principles.**

(iii) **Study Director (SD): the SD is the only point of control of a study and the only one who supports the study, since they are responsible for the study from the beginning to the end.**

The regulatory authority can evaluate the data from a study in two ways:

(i)   Repeating all experiments, or

(ii) Rebuilding, step-by-step, all activities and circumstances that led to the outcome of the study.

Primarily, the GLP was been developed to combatfraud in the generation and reporting of safety data. However, the goal of the GLP is much broader, as it is not only a control mechanism allowing regulatory agencies to judge the integrity of a study. The principles of the GLP are designed as a tool that also allows improvements in the study and data quality, by applying strict requirements regarding documentation, providing the ability to rebuild any activity, and making the way back to its inception. The GLP requirements are related to several issues and are aimed at different organizational levels of a test facility. Many of these requirements can be considered "common sense" and should be followed when working under the principles of GLP. Broadly, the requirements can be summarized in three ways that are the central ideas in GLP:

(i)   *Reproducibility*: This provides assurance that the experiments were conducted as described in the report submitted to the regulatory agency.

(ii) *Responsibility*: this is closely related to the first term. The documentation required to conduct a study, according to the principles of GLP, will report who did what and who can be held responsible for likely errors.

(iii) *Awareness*: the principles of GLP raise awareness of broad tasks, such as administrative work, which is aimed at the quality and transparency of studies conducted in the test facility.

Study of Distribution, Metabolism and Pharmacokinetics (DMPK) of New Substances

The main characteristics that determine the success of a drug candidate are directly related to its kinetic properties. In this context, through non-clinical studies that are planned and properly executed, it is possible to characterize the pharmacokinetic profile of a substance aiming to establish an adequate dosage that enables patient adherence to treatment and the correct interpretation of results obtained from efficacy and safety studies.

Mostly, the toxic effects of substances leading to the discontinuation of their development are associated with prolonged systemic drug exposure, the formation of toxic/reactive metabolites and/or possible drug-drug interactions. A number of drugs such as troglitazone, trovafloxacin and bromfenac (oral formulation) were withdrawn from the market due to the formation of reactive hepatotoxic metabolites and other drugs that are still on the market such as acctaminophen and amiodarone, etc., Can cause hepatotoxicity when used at high doses.

Considering the idea that the pharmacokinetic properties of a substance are determinant to its success in clinical studies, the pharmaceutical industry has introduced DMPK (Drug Metabolism and Pharmacokinetic) studies in the early phases of new drug development. A recent study conducted by four important pharmaceutical companies (AstraZeneca, UK; Eli Lilly and Company, USA; GlaxoSmithKline, UK; and Pfizer, USA) showed that, with the introduction of DMPK studies during the lead identification phase, the pharmacokinetics represent only 5% of the reasons for failure in clinical development. From the financial viewpoint, the discontinuity of a project during the final phases generates losses of around 90% of the total investment. Therefore, the DMPK studies carried out during the non-clinical phases are extremely important since they reduce the time and costs expended with the development of new drugs. Additionally, the DMPK performed during the early phases of drug development provide important information about the structure of a molecule and possible modifications that can be made to optimize its DMPK properties.

Finally, data on the pharmacokinetic properties of a new substance together with preliminary studies about its safety and efficacy are critical to take the decision to continue or not (go/no-go decision) the development process of a new drug.

Desirable DMPK Properties of a Substance (Candidate) Intended for Oral Administration

A substance intended to be used via the oral route should present some critical properties in DMPK studies:

(i)   Water solubility;

(ii) High permeability and low efflux in caco-2 cells;

(iii) Sufficient bioavailability to reach the desirable plasma and organ to produce the pharmacological effect;

(iv) Adequate half-life time to the intended posology scheme in human;

(v)   Linear pka;

(vi) Elimination that is not dependent on a single route or on a single metabolization enzyme, without forming active or reactive metabolites in large amounts and without interacting with metabolization enzymes in relevant concentrations;

(vii) Acceptable safety margin (therapeutic index, preferably higher than 10 times)

(viii) Established pk-pd (pharmacodynamics) relation

Toxicokinetic Assay

Toxicokinetic assays are usually performed during toxicology studies and should be conducted in accordance with the GLP rules. Toxicokinetic assays measure the systemic exposure of a substance in animals and establish the relationship between the dose administered and the time course of the substance in the toxicity studies. Indeed, the PK profile determination of a substance following administration of multiple doses enables the best interpretation of the toxicological findings. The toxicokinetic study also evaluates the potential of a substance to accumulate in a specific organ and/or tissue. Thus, the data generated with these studies should contribute to the data obtained with the toxicology studies in terms of interpretation of the toxicity tests and in comparison, with the clinical data as part of the risk and safety evaluation in humans. Toxicokinetic studies are part of the non-clinical test battery recommended by the regulatory agencies

Two protocols can be applied to toxicokinetic studies:

(i) A full protocol that answers all aforementioned questions or

(ii) A reduced protocol, in which only the main questions are answered to corroborate the interpretation of the toxicology findings.

In the full toxicokinetic protocol, other biological matrices should be collected besides blood, such as excrements (urine and faeces), fat, muscles, liver, kidneys, possible target-organs and skin (when the substance test is administrated by the dermal route). Moreover, if the substance in its parent form and/or its metabolites is volatile, additional animal groups should be included for collecting excrement, carcasses and expired air to determine the extension of absorption and the route(s) of elimination of the substance. The design of the study and the selection of the experimental protocol should be defined on a case-by-case basis; overall, they should be able to provide enough information to evaluate the risks and safety of the candidate substance. Considering the importance of evaluating PK properties during the development of new drugs, the assays described in **Table 3.1** are highly recommended.

Table 3.1 Recommended non-clinical assays of ADME/PK.

Non-GPL		GLP
In vitro	*In vivo*	*In vivo*
Physical/chemical properties [lipophilicity (log P/log D), solubility, chemical stability (pKa)]	Pharmacokinetic profile (concentration versus time) - Area under the curve - Cmax - Tmax - Distribution - Clearance - Half-life time	Toxicokinetic - Pharmacokinetic profile (concentration versus time) - Area under the curve - Cmax - Tmax - Distribution - Clearance - Half-life time
Metabolic stability	Biodispersibility/bioavailability	Biodispersibility
Hepatic clearance	Linearity	Metabolization
Interactions between substances (inhibition/ induction of CYPs)	Metabolization	Route of excretion
Physiological characteristics (plasma protein/tissue binding)	Route of excretion	Quantification of biological fluids, organs, tissues, excrements and expired air (when necessary
Permeability		
Plasmatic stability and total blood/plasma partition		

The performance of non-clinical tests in sequence is an important factor in the development of a new medicine. Despite of there is no standard program to execute the assays, a well-designed plan helps avoid several errors or unnecessary tests, besides saving time and financial resources. In Figure 3.4, the nonclinical studies that should be conducted during drug development Thus, in this section, we will discuss the main non-clinical safety tests that are required in the process of drug development, such as mutagenicity tests, acute, sub-chronic and chronic toxicity, developmental and reproductive toxicity, carcinogenicity, local tolerance and safety pharmacology.

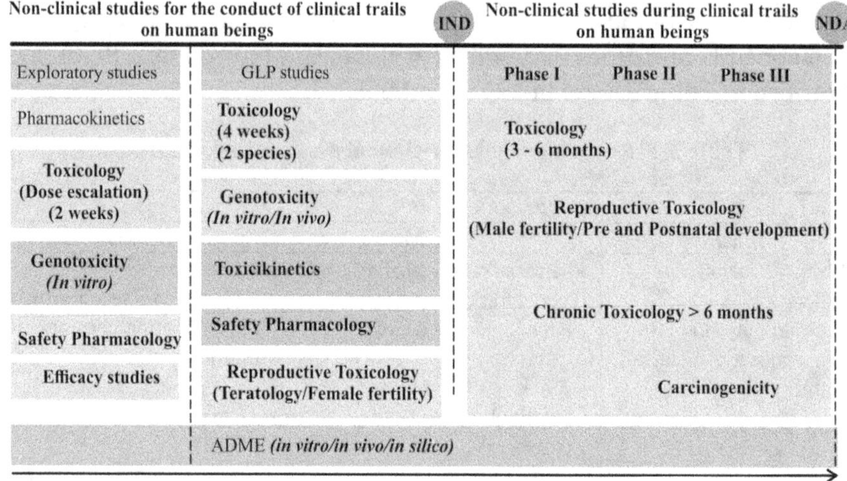

Fig. 3.4 Steps of non-clinical studies in the drug development process GLP: Good laboratory practice; IND: Investigational new drug; NDA: New drug application; ADME: absorption, distribution, metabolism and elimination

Investigational New Drug (IND) Application

An application has been filed with the FDA to start clinical trials in humans if the drug was found to be safe from the reports of preclinical trials. The IND application should provide high-quality preclinical data to justify the testing of the drug in humans. Almost 85% of drugs are subjected to clinical trials, for which IND applications are filed. A firm or institution called a Sponsor, is responsible for submitting the IND application.

A pre-IND meeting can be arranged with the FDA to discuss several issues:

- The design of animal research, is required to support the clinical studies

- The intended protocol for conducting the clinical trial

- The chemistry, manufacturing, and control of the investigational drug.

Such a meeting will help the Sponsor organize animal research, gather data and design the clinical protocol based on suggestions by the FDA.

The next step is phase-I clinical trials (1–3 years) on human subjects (~100). The drug's safety profile and pharmacokinetics of the drug are focused on this phase. Phase-II trials (2 years) are performed if the drug successfully passes phase-I. To evaluate dosage, broad efficacy and additional safety in people (~300) are the main objectives of the phase-II. If evidence of effectiveness is shown in phase II, phase III studies (3–4 years) begins. This phase-III concerns more about the safety and effectiveness of drug from data of different populations, dosages and its combination with other drugs in several hundred to about 3,000 peoples.

Drug Approval Process in United States

In 1820, the new era of USA drug regulation was started with the establishment of U.S. Pharmacopoeia. In 1906, Congress passed the original Food and Drugs Act, which required that drugs must meet official standards of strength and purity. However, in 1937, due to the sulfanilamide tragedy, the Federal Food, Drug and Cosmetic Act (of 1938) was enacted and added new provisions that new drugs must be shown safe before marketing. Further, in 1962, the Kefauver-Harris Amendment Act was passed which require that manufacturers must prove that the drug is safe and effective (for the claims made in labelling).

The United States has perhaps the world's most stringent standards for approving new drugs. Drug approval standards in the United States are considered by many to be the most demanding in the world. The Food and Drug Administration (FDA) is responsible for protecting and promoting public health. Like general drug approval process, FDA's new drug approval process is also accomplished in two phases: clinical trials (CT) and a new drug application (NDA) approval. The FDA approval process begins only after the submission of an investigational new drug (IND) application.

Types of Investigational New Drugs (IND)

- An Investigator IND is submitted by a physician who both initiates and conducts an investigation, and under whose immediate direction the investigational drug is administered or dispensed. A physician might submit a research IND to propose studying an unapproved drug, or an approved product for a new indication or in a new patient population.

- Emergency Use IND allows the FDA to authorize use of an experimental drug in an emergency situation that does not allow time for submission of an IND in accordance with 21CFR, Sec. 312.23 or Sec. 312.20. It is also used for patients who do not meet the criteria of an existing study protocol, or if an approved study protocol does not exist.

- Treatment IND is submitted for experimental drugs showing promise in clinical testing for serious or immediately life-threatening conditions while the final clinical study is conducted and the FDA review takes place.

Categories and Areas of Investigational New Drug (IND)

There are two IND categories:

- Commercial

- Research (non-commercial)

The IND application must contain information in three broad areas:

- Animal Pharmacology and Toxicology Studies - Preclinical data to permit an assessment of whether the product is reasonably safe for initial testing in humans. Also included are any previous experience with the drug in humans (often foreign use).

- Manufacturing Information - Information about the composition, manufacturer, stability, and controls used for manufacturing the drug substance and the drug product. This information is assessed to ensure that the company can adequately produce and supply consistent batches of the drug.

- Clinical Protocols and Investigator Information - Detailed protocols for proposed clinical studies to assess whether the initial-phase trials will expose subjects to unnecessary risks. Also, information on the qualifications of clinical investigator professionals (generally physicians) who oversee the administration of the experimental compound to assess whether they are qualified to fulfill their clinical trial duties. Finally, commitments to obtain informed consent from the research subjects, to obtain a review of the study by an institutional review board (IRB) and to adhere to the investigational new drug regulations.

Once the IND is submitted, the sponsor must wait 30 calendar days before initiating any clinical trials. During this time, FDA has an opportunity to review the IND for safety to assure that research subjects will not be subjected to unreasonable risk.

Ivestigator's Brochure

The Investigator's Brochure (IB) is a compilation of the clinical and nonclinical data on the investigational product(s) that are relevant to the study of the product(s) in human subjects. Its purpose is to provide the investigators and others involved in the trial with the information to facilitate their understanding of the rationale for, and their compliance with, many key features of the protocol, such as the dose, dose frequency/interval, methods of administration, and safety monitoring procedures. The IB also provides insight to support the clinical management of the study subjects during the course of the clinical trial. The information should be presented in a concise, simple, objective, balanced, and non-promotional form that enables a clinician, or potential investigator, to understand it and make his/her own unbiased risk-benefit assessment of the appropriateness of the proposed trial. For this reason, a medically qualified person should generally participate in the editing of an IB, but the contents of the IB should be approved by the disciplines that generated the described data.

Generally, the sponsor is responsible for ensuring that an up-to-date IB is made available to the investigator(s) and the investigators are responsible for providing the up-to-date IB to the responsible IRBs/IECs. In the case of an investigator- sponsored trial, the sponsor investigator should determine whether a brochure is available from the commercial manufacturer. If the investigational product is provided by the sponsor-investigator, he or she should provide the necessary information to the trial personnel. In cases where the preparation of a formal IB is impractical, the sponsor-investigator should provide, as a substitute, an expanded background information section in the trial protocol that contains the minimum current information described in this guidance.

Contents of the Investigator's Brochure

The IB should contain the following sections, each with literature references where appropriate:

- **Title Page:** This should provide the sponsor's name, the identity of each investigational product (i.e., research number, chemical or approved generic name, and trade name(s) where legally permissible and desired by the sponsor), and the release date. It is also suggested that an edition number, and a reference to the number and date of the editions it supersedes, be provided.

- **Confidentiality Statement:** The sponsor may wish to include a statement instructing the investigator/recipients to treat the IB as a confidential document for the sole information and use of the investigator's team and the IRB/IEC.

- **Table of Contents**

- **Summary:** A brief summary (preferably not exceeding two pages) should be given, highlighting the significant physical, chemical, pharmaceutical, pharmacological, toxicological, pharmacokinetic, metabolic, and clinical information available that is relevant to the stage of clinical development of the investigational product.

- **Introduction:** A brief introductory statement should be provided that contains the chemical name (and generic and trade name(s) when approved) of the investigational product(s), all active ingredients, the investigational product(s) pharmacological class and its expected position within this class (e.g., advantages), the rationale for performing research with the investigational product(s), and the anticipated prophylactic, therapeutic, or diagnostic indication(s). Finally, the introductory statement should provide the general approach to be followed in evaluating the investigational product.

- **Physical, Chemical, and Pharmaceutical Properties and Formulation:** A description should be provided of the investigational product substance(s) (including the chemical and/or structural formula (e)), and a summary should be provided of the relevant physical, chemical, and pharmaceutical properties.

- **Nonclinical Studies:** The results of all relevant nonclinical pharmacology, toxicology, pharmacokinetic, and investigational product metabolism studies should be provided in summary form. This summary should address the methodology used, the results, and a discussion of the relevance of the findings to the investigated therapeutic and the possible unfavorable and unintended effects in humans.

The following sections should discuss the most important findings from the studies, including the dose response of observed effects, the relevance to humans, and any aspects to be studied in humans. If applicable, the effective and nontoxic dose findings in the same animal species should be compared (i.e., The therapeutic index should be discussed). The relevance of this information to the proposed human dosing should be addressed. Whenever possible, comparisons should be made in terms of blood/tissue levels rather than on a mg/kg basis.

(a) *Nonclinical Pharmacology:* A summary of the pharmacological aspects of the investigational product and, where appropriate, its significant metabolites studied in animals should be included. Such a summary should incorporate studies that assess potential therapeutic activity (e.g., efficacy models, receptor binding, and specificity) as well as those that assess safety (e.g., Special studies to assess pharmacological actions other than the intended therapeutic effect(s)).

(b) ***Pharmacokinetics and Product Metabolism in Animals:*** A summary of the pharmacokinetics and biological transformation and disposition of the investigational product in all species studied should be given. The discussion of the findings should address the absorption and the local and systemic bioavailability of the investigational product and its metabolites, and their relationship to the pharmacological and toxicological findings in animal species.

(c) ***Toxicology:*** A summary of the toxicological effects found in relevant studies conducted in different animal species should be described under the following headings where appropriate:
  (i)   Single dose;
  (ii) Repeated dose;
  (iii) Carcinogenicity;
  (iv) Special studies (e.g., irritancy and sensitization);
  (v)   Reproductive toxicity;
  (vi) Genotoxicity (mutagenicity).

Effects in Humans: A thorough discussion of the known effects of the investigational product(s) in humans should be provided, including information on pharmacokinetics, metabolism, pharmacodynamics, dose response, safety, efficacy, and other pharmacological activities. Where possible, a summary of each completed clinical trial should be provided. Information should also be provided regarding results from any use of the investigational product(s) other than in clinical trials, such as from experience during marketing.

(a) ***Pharmacokinetics and Metabolism of drug in Humans***
A summary of information on the pharmacokinetics of the investigational product(s) should be presented, including the following, if available:
  (i)   Pharmacokinetics (including metabolism, as appropriate, and absorption, plasma protein binding, distribution, and elimination).
  (ii) Bioavailability of the investigational product (absolute, where possible, and/or relative) using a reference dosage form.
  (iii) Population subgroups (e.g., gender, age, and impaired organ function).
  (iv) Interactions (e.g., product-product interactions and effects of food).
  (v)   Other pharmacokinetic data (e.g., results of population studies performed within clinical trial(s)).

(b) *Safety and Efficacy*: A summary of information should be provided about the investigational products/products' (including metabolites, where appropriate) safety, pharmacodynamics, efficacy, and dose response that were obtained from preceding trials in humans (healthy volunteers and/or patients). In cases where several clinical trials have been completed, the use of summaries of safety and efficacy across multiple trials by indications in subgroups may provide a clear presentation of the data. Tabular summaries of adverse drug reactions for all clinical trials (including those for all studied indications) would be useful.

The IB should provide a description of the possible risks and adverse drug reactions anticipated on the basis of prior experiences with the product under investigation and with related products. A description should also be provided of the precautions or special monitoring to be done as part of the investigational use of the product(s).

(c) *Marketing Experience*: The IB should identify countries where the investigational product has been marketed or approved. Any significant information arising from the marketed use should be summarized (e.g. Formulations, dosages, routes of administration, and adverse product reactions). The IB should also identify all the countries where the investigational product did not receive approval/registration for marketing or was withdrawn from marketing/registration.

- **Summary of Data and Guidance for the Investigator:** This section should provide an overall discussion of the nonclinical and clinical data, and should summarize the information from various sources in different aspects of the investigational product(s), wherever possible. In this way, the investigator can be provided with the most informative interpretation of the available data and with an assessment of the implications of the information for future clinical trials.

New Drug Application (NDA)

A new drug application (NDA) can be filed only when the drug successfully passes all three phases of clinical trials and includes all animal and human data, data analyses, pharmacokinetics of drug and its manufacturing and proposed labelling. The preclinical, clinical reports and risk-benefit analysis (product's beneficial effects outweigh its possible harmful effects) are reviewed at the Centre for Drug Evaluation and Research by a team of scientists. If clinical studies confirm that a new drug is relatively safe and effective, and will not pose unreasonable risks to patients, the manufacturer files a New Drug Application (NDA), the actual request to manufacture and sell the drug in the United States. Figure 3 describes the drug approval process in USA.

Drug Approval Process in India

The Drug and Cosmetic Act 1940 and Rules 1945 were passed by India's parliament to regulate the import, manufacture, distribution and sale of drugs and cosmetics. The Central Drugs Standard Control Organization (CDSCO) and the office of its leader, the Drugs Controller General (India) [DCGI] were established. In 1988, the Indian government added Schedule Y to the Drug and Cosmetics Rules 1945. Schedule Y provides the guidelines and requirements for clinical trials, which was further revised in 2005 to bring it at par with internationally accepted procedure.

When a company in India wants to manufacture/import a new drug it has to apply to seek permission from the licensing authority (DCGI) by filing in Form 44 and submitting the data as given in Schedule Y of Drugs and Cosmetics Act 1940 and Rules 1945. To prove its efficacy and safety in the Indian population it has to conduct clinical trials in accordance with the guidelines a specified in Schedule Y and submit the report of such clinical trials in a specified format.

Rule- 122A of the Drug and Cosmetics Act says that the clinical trials may be waived in the case of new drugs which are approved and being used for several years in other countries.

Section 2.4 (a) of Schedule Y of Drugs and Cosmetics Act 1940 and Rules of 1945 says that for those drug substances which are discovered in India all phases of clinical trials are required.

Section 2.4 (b) of Schedule Y of Drugs and Cosmetics Act 1940 and Rules 1945 says that for those drug substances which are discovered in countries other than India; the applicant should submit the data available from other countries and the licensing authority may require him to repeat all the studies or permit him to proceed from Phase III clinical trials.

Demonstration of safety and efficacy of the drug product for use in humans is essential before the drug product can be approved for import or manufacturing of new drug by the applicant by the Central Drugs Standard Control Organization (CDSCO). The regulations under Drugs and Cosmetics Act 1940 and its rules 1945, 122A, 122B, and 122D and further Appendix I, IA and VI of Schedule Y, describe the information required for the approval of an application to import or manufacture of new drug for marketing.

The changes in the Drugs and Cosmetics Act includes, establishing definitions for Phase I-IV trials and clear responsibilities for investigators and sponsors. After the clinical trials were further divided into two categories in 2006. In one category (category A) clinical trials can be conducted in other markets with competent and mature regulatory systems, whereas the remaining ones fall in to another category (category B) Other than A. Clinical trials of category A (approved in the U.S., Britain, Switzerland, Australia, Canada, Germany, South Africa, Japan, and European Union) are eligible for fast tracking in India and are likely to be approved within eight weeks. The clinical trials of category B are under more scrutiny, and are approved within 16 to 18 weeks.

An application to conduct clinical trials in India should be submitted along with the data of chemistry, manufacturing, control and animal studies to DCGI. The date regarding the trial protocol, investigator's brochures, and informed consent documents should also be attached. A copy of the application must be submitted to the ethical committee and the clinical trials are conducted only after approval of DCGI and ethical committee.

To determine the maximum-tolerated dose in humans, adverse reactions, etc. On healthy human volunteers, Phase I clinical trials are conducted. The therapeutic uses and effective dose ranges are determined in Phase II trials in 10–12 patients at each dose level. The confirmatory trials (Phase III) are conducted to generate data regarding the efficacy and safety of the drug in ~ 100 patients (in 3–4 centres) to confirm efficacy and safety claims. Phase III trials should be conducted on a minimum of 500 patients spread across 10–15 centres, if the new drug substance is not marketed in any other country.

The new drug registration (using form # 44 along with full pre-clinical and clinical testing information) is applied after the completion of clinical trials. Comprehensive information on the marketing status of the drug in other countries is also required, other than the information on safety and efficacy. Information regarding the prescription, samples and testing protocols, product monograph, labels, and cartons must also be submitted.

The application can be reviewed in a range of about 12–18 months. After the NDA approval, when a company is allowed to distribute and market the product, it is considered as in Phase IV trials, in which new uses or new populations, long-term effects, etc., are explored.

Through the International Conference on Harmonization (ICH) process, the Common Technical Document (CTD) guidelines have been developed for Japan, European Union, and the United States.

Most countries have adopted the CTD format. Hence, CDSCO has also decided to adopt the CTD format for technical requirements for registration of pharmaceutical products for human use.

Stages of Approval

1. Submission of Clinical Trial application to evaluate safety and efficacy.

2. Requirements for permission of new drug approval.

3. Post approval changes in biological products: quality, safety and efficacy documents.

4. Preparation of the quality information for drug submission for new drug approval.

Clinical Research and Bioequivalence Studies

The earliest clinical research is traced back to as early as Judeo-Christian and Eastern civilizations. There is also evidence that ancient Chinese medicine included clinical studies. Hippocrates, the father of modern medicine, is considered the first clinical researcher for his disciplined investigations. Clinical research has been greatly influenced by the Greek and Roman researchers of their times. They laid the foundations of modern-day science. During the middle ages and the Renaissance, the improvements in medicine became evident and the infrastructure for clinical research began to develop. The Renaissance (1453–1600) represented the revival of learning and transition from medieval to modern conditions; many great clinicians and scientists prospered. Studies of blood began in the 17th century. The 18th century brought extraordinary advances in the biological sciences and medicine. In the 20th century, clinical research began to flourish with the help of well-established medical colleges. Two other dominant drivers of the progress in medicine through clinical research are government investment in biomedical research and private investment in the pharmaceutical industry.

Objective of the Guideline

This guideline describes the principles of procedures of bioequivalence studies of generic products. The objective of the study is to assure therapeutic equivalence of generic products to innovator products. In the bioequivalence study, bioavailability should be compared for innovator and generic products. If this is not feasible, pharmacological effects supporting efficacy or therapeutic effectiveness in major indications should be compared (These comparative tests are hereafter called pharmacodynamic studies and clinical studies, respectively). For oral drug products, dissolution tests should be performed, since they provide important information concerning bioequivalence.

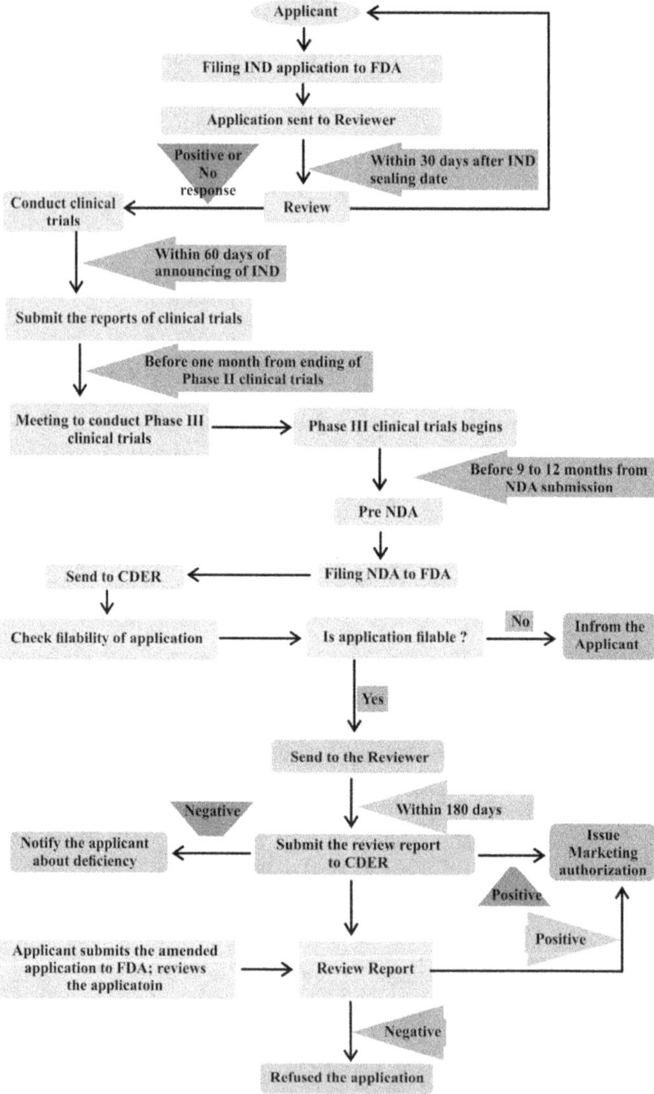

Fig. 3.5 Drug approval process in USA

Guideline for Bioequivalence Studies of Generic Products

(Oral Conventional Dosage Forms and Enteric Coated Products)

I. Reference and test products

Dissolution tests should be performed using 6 units or more for three lots of an innovator product by the paddle method at 50 rpm. Among these, the one with intermediate dissolution under the most discriminative condition, where the difference in dissolution between the fastest and slowest lots is the largest, should be selected as the reference product. If the dosage form is an aqueous solution or the drug is administered as an aqueous solution, any innovator product can be used as the reference product.

As the test product of generic drugs, it is recommended to use a lot of industrial scale. However, a lot of 1/10 or larger of industrial scale can also be used as the test product, which should be the same as the production lots in the manufacturing method, quality and bioavailability. The drug content or potency of the reference product should be close to the label claim, and the difference in drug content or potency between test and reference products should be less than 5%.

II. Bioequivalence studies

A. Test methods

An appropriate study protocol, including the required number of subjects and sampling intervals should be determined according to preliminary studies and the previously reported data. The rationale of the protocol should also be described.

1. Design

In principle, crossover studies should be employed with random assignment of individual subjects to each group. Parallel designs can be employed for drugs with extremely long half-lives. Replicate crossover studies may be useful for drugs with large intra-subject variability in clearance.

2. Number of subjects

A sufficient number of subjects for assessing bioequivalence should be included. If bioequivalence cannot be demonstrated because of an insufficient number, an add-on subject study can be performed using not less than half the number of subjects in the initial study. A sample size of 20 (n=10/group) for the initial study and a pooled size of 30 for initial plus add-on subject study may suffice if test and reference products are equivalent in dissolution and similar in average AUC and C_{max} as described under Sec.3 A.II.2.5. Multiple dose studies or studies with stable isotopes may be useful for highly variable drugs that require large sample sizes.

3. Selection of subjects

In principle, healthy adult volunteers should be employed. When the test and reference products showed a significant difference in dissolution at around pH 6.8 by the dissolution test or between pH 3 and 6.8 for products containing basic drugs, subjects with low gastric acidity (achlorhydric subjects) should be employed unless the application of the drug is limited to a specific population. This rule is not applied to enteric coated products. If the use of the drug is limited to a special population and test and reference products show a significant difference in dissolution even under the conditions of the dissolution test, the in vivo test should be performed using subjects from the target population.

4. Drug administration

a) **Dose:** One dose unit or a clinical usual dose should generally be employed. A higher dose, which does not exceed the maximal dose of the dosage regimen, may be employed when analytical difficulties exist, such as a high detection limit.

b) **Single vs. Multiple - dose studies:** In principle, bioequivalence studies should be performed by single dose studies. Multiple dose studies may be employed for drugs which are repeatedly administered to patients.

  (i) **Single dose studies:** Drugs are usually given to subjects with 100-200 ml water(normally 150 ml) after fasting for more than 10 hr. Fasting lasts for at least 4 hrpost-dose. If the postprandial dose is specified in the dosage regimen, or if the bioavailability in fasting state is very poor, or high incidence of severe adverse events is anticipated, drugs may be given after food. In the fed study, a low-fat diet of 700 kcal or less containing not more than 20 % by energy of the lipid should be employed. The meal should be eaten within 20 min, and drugs are administered according to the dosingregimen or 30 min after the meal, if the dosing time is not indicated in the regimen.

  (ii) **Multiple dose studies:** Drugs should, in principle, be administered to subjects underfasting conditions as in the single dose studies when biological fluids are sampled for the assessment of bioavailability. In the time period before fluids are sampled, drugs should repeatedly be given between meals (drugs should be administered more than 2 hr after a meal) at constant intervals. `

5. Measurement of drug substances

    a) Biological fluids to be sampled: Blood samples should generally be employed. Urine samples may be used if there is a rationale.

    b) Sampling schedule: Blood samples should be taken at a frequency sufficient for assessing C_{max}, AUC and other parameters. Sampling points should be at least 7, including zero time, 1 point before C_{max}, 2 points around C_{max} and 3 points during the elimination phase. Sampling should be continued until AUC is above 80% of AUC (normally more than 3 times the elimination half-life after t_{max}). However, when the elimination half-life of the parent drug or active metabolites is extremely long, blood samples should be collected for at least 72 hr. When urine samples are used, they should be collected in the same manner as blood samples. If F can be determined by deconvolution, such extended sampling may not be required, although the sampling should be continued until drug absorption is complete.

    c) Drug substances to be measured: Parent drugs should, in principle, be measured. Major active metabolites may be measured instead of the parent drug, if it is rational. Stereoselective assay is not generally required. However, when it is indicated that there exist stereo-isomers with different activities for the main pharmacological effect, and stereoselective absorption or elimination, dependent on the absorption rate is noticeable, the enantiomer with higher activity should be measured.

    d) Assay: Analytical methods should be fully validated regarding specificity, accuracy, precision, linearity, quantification limit, and stability of substances in samples and so forth.

6. Washout periods

Washout periods in crossover studies between the administration of test and reference products should usually be more than 5 times the elimination half-life of the parent drug or active metabolites.

B. Assessment of bioequivalence

1. Bioequivalence range

The acceptable range of bioequivalence is generally 0.8–1.25 as the ratios of the average AUC and C_{max} of the test product to the reference product, when the parameters are logarithmically distributed. The acceptable range is generally -0.2 - +0.2 as the ratio of the relative difference in the mean AUC and C_{max} between reference and test products to those of the reference product, when the parameters are normally distributed. For drugs with pharmacologically mild actions, a wider range may be acceptable. The acceptable ranges for other parameters, such as t_{max} should be determined for each drug.

2. Parameters to be assessed

When blood samples are used, AUC^t and C_{max} should be subjected to the bioequivalence assessment in single dose studies and AUC^t and C_{max} in multiple dose studies, C_{max} is an observed value and AUC is calculated using the trapezoidal integration method. If F can be estimated by deconvolution, F can be used for AUC.

Parameters such as AUC^8, t_{max}, MRT, and k_{el} should be submitted as reference data. For multiple dose studies, Ct is also used as a reference parameter. When urine samples are used, A^{et}, U_{max} and U_t are employed instead of AUC^t, AUC C_{max} and C_t.

3. Logarithmic transformation

Pharmacokinetic parameters, except for t_{max} should in principle be statistically analysed after logarithmic transformation.

4. Statistical analysis

A 90% shortest confidence interval or two one-sided t tests with the significance level of 5% should be used. Other reasonable statistical methods can also be used. When the add-on subject study is performed and there are no fundamental differences between the two studies in formulation, design, assay and subjects, data from the initial and add-on subject studies can be pooled and statistically analysed. In the pooled analysis, the study must be added as a source of variation.

5. Acceptance criteria

Products are considered bioequivalent, if the 90% confidence interval of the difference in the average values of logarithmic AUC and Cmax between test and reference products is within the acceptable range of log (0.8) – log (1.25). However, even though the confidence interval is not in the above range, test products are accepted as bioequivalent, if the following three conditions are satisfied; 1) the total sample size of the initial bioequivalence study is not less than 20 (n=10/group) or the pooled sample size of the initial and add-on subject studies is not less than 30, 2) the differences in average values of logarithmic AUC and C_{max} between two products are between log (0.9) – log (1.11), and 3) dissolution rates of test and reference products are evaluated to be equivalent under all dissolution testing conditions. However, the 3rd rule cannot be applied to slowly dissolving products from which more than 80% of a drug does not dissolve within the final testing time (2 hr in pH 1.2 medium and 6 hr in others) under any conditions of the dissolution tests. Reference parameters should be subjected to statistical assessment. If a significant difference is detected in the parameters between reference and test products, effects of this difference on therapeutic equivalence should be explained.

C. Pharmacodynamic studies

These studies are performed to establish the equivalence of products using pharmacological activity in humans as an index. This is applied to pharmaceuticals which do not produce measurable concentrations of the parent drug or active metabolite in blood or urine and those whose bioavailability does not reflect therapeutic effectiveness. In the study, it is desirable to compare the efficacy-time profiles. For antacids or digestive enzymes, suitable in vitro efficacy tests can be employed. The Acceptance criteria of equivalence in this study should be established by considering the pharmacological activity of each drug.

D. Clinical studies

These studies were performed to establish the equivalence of drugs using clinical effectiveness as an index. If bioequivalence and pharmacodynamic studies are impossible or inappropriate, this study is applied.

The Acceptance criteria of equivalence in this study should be established by considering the pharmacological characteristics and activity of each drug.

E. Reporting of test results

a) Samples

(i) Brand names and lot No. of the reference product. Code No. or name, lot No. and lot sizes of test products

(ii) Type of dosage form

(iii) Name of drug substances

(iv) Labelled contents or potencies

(v) Measured contents or potencies and assay procedures

(vi) Solubility of drugs at different pH and in water for dissolution tests

(vii) Particle size or specific surface area for low solubility drugs and their measurement procedures

(viii) Types of polymorphs and solubility

(ix) Others (for example, p^{ka} and physicochemical stability)

b) Results of tests

(i) Summary

(ii) Dissolution tests:

(a) List of test conditions (apparatus, stirring speed, types and volumes of test solutions)

(b) Assay: Methods and summary of validation

(c) Summary of the validation of the dissolution tests

(d) Results

(a) *Results of preliminary tests performed to select a reference product.*

Tables listing dissolution percent of individual samples under each testing condition, average values and standard deviations of each group.

Figures comparing the average dissolution curves of each lot under each testing condition

(b) *Results of preliminary tests were performed to select test media.*

(c) *Comparison* of reference and test products

(iii) Tables list dissolved amounts of individual samples under each testing condition,

(iv) The average values and standard deviations of the test and reference products.

(v) Figures comparing the average dissolution curves of test and reference products under

(vi) Each testing conditions.

c) Bioequivalence studies

The following should be described. Preliminary test items should also be reported.

i) Experimental conditions

(a) Subjects: Age, sex, body weight and other data obtained from laboratory tests are described. Individual gastric acidity should be reported if necessary or otherwise available.

(b) Drug administration fasting time, co-administered water volume, and times of drug administration and food ingestion are described. In the case of postprandial administration, menu and content of meal (protein, fat, carbohydrate, calories and others), and times of food ingestion during studies.

(c) Assay: procedure and summary of validation.

(ii) Results

    a) Individual subject data

Tables showing drug levels in biological fluids at each sampling time, C_{max}, C_t, AUC^t, AUC^t, AUC k_{el}, t_{max} and MRT. The correlation coefficient for determining k_{el} should be reported together with the time points used.

The ratios of Cmax and AUC^t of the test product to those of the reference product for each individual should be reported.

Figures comparing individual drug level-time profiles of the two products drawn on a linear/linear scale.

    b) Averages and standard deviations

Tables showing averages and standard deviations of raw data of drug levels in biological fluids at each time point, C_{max}, C_t, AUC^t, AUC^t, AUC k_{el}, t_{max} and MRT.

The ratios of the average of Cmax and AUC^t of the test product to those of the reference product should be reported.

Figures comparing average drug level-time profiles of the two products drawn on a linear/linear scale.

    c) Statistical analysis and equivalence assessment

Analysis of variance tables for C_{max}, C_t, AUC^t, AUC^t, AUC k_{el}, t_{max} and MRT which are logarithmically transformed when required. The 90% confidence intervals for C_{max} and AUC_t or alternative statistical results. For other parameters, statistical testing results of the null-hypotheses should be reported where the average values of test and reference products are assumed to be equivalent.

    d) Analysis of pharmacokinetic parameters

If deconvolution is used, the program, algorithm, pharmacokinetic models and fitting information should be listed.

    e) Others

Information on dropouts (data, reasons), monitoring records of health of subjects.

(d) Pharmacodynamic studies

Reporting of results should follow that of bioequivalence studies.

(e) Clinical studies

Reporting of results should follow that of bioequivalence studies.

Clinical Research Protocols

Every clinical investigation begins with the development of a clinical protocol. The protocol is a document that describes how a clinical trial will be conducted (the objective(s), design, methodology, statistical considerations and organization of a clinical trial,) and ensures the safety of the trial subjects and integrity of the data collected. A research protocol is a document that describes the background, rationale, objectives, design, methodology, statistical considerations, and organization of a clinical research project.

Clinical research is conducted according to a plan (a protocol) or an action plan. The protocol demonstrates the guidelines for conducting a trial. It illustrates what will be made in the study by explaining each essential part of it and how it is carried out. It also describes the eligibility of the participants, the length of the study, the medications and the related tests.

The protocol is directed by a chief researcher. The health of the participants' will be regularly checked by members of the research team to ultimately ensure the study's safety and effectiveness.

Purpose of the Research Proposal

- To raise the question to be researched and to clarify its importance.

- To collect existing knowledge and discuss the efforts of other researchers who have worked on the related questions (Literature review).

- To develop a hypothesis and objectives.

- To clarify ethical considerations.

- To suggest the methodology required for solving the question and achieving the objectives.

- To discuss the requirements and limitations for achieving the objectives.

Benefits of the Proposal for a Researcher

- Allows the researchers to plan and review the project's steps.

- Serves as a guide throughout the research.

- Forces time and budget estimates

What is a protocol review?

Clinical trials must be approved and monitored by an Institutional Review Board that ensures that the risks are negligible and are worth any potential benefits. It is an independent committee that consists of physicians, dentists, statisticians, and members of the community. The committee ensures that clinical trials are ethical and that the rights of all participants are protected. The board must initially approve and periodically review the research.

Components of a Research Protocol

The topics that should be covered in the protocol are

  (i) Title of the study

  (ii) Administrative details

  (iii) Project summary

  (iv) Introduction to the research topic, background (Literature review)

  (v) Preliminary studies

  (vi) Study objectives and/or questions. Statement of the problem.

  (vii) Methodology: Study design, study population and methods of recruitment, variable list, sample size, methods of data collection, data collection tools, plan of analysis (analysis of data)

  (viii) Project management: Work plan (Timeline - proposed schedule)

  (ix) Strengths and limitations of the study

  (x) Issues for ethical review and approval

  (i) **Title of the study:** Title of proposal should be accurate, short, concise and identify. **What** is the study about, **who** are the targets, **where** is the setting of the study and **when** it is launched, if applicable- It should make the main objective clear, convey the main purpose of the research and mention the target population? Carry maximum information about the topic in a few words; it is a good practice to keep the title to within 12–15 words. It should convey the idea about the area of research and what methods are going to be used in a compact, relevant, accurate, attractive, easy to understand, and informative way.

  (ii) **Administrative Details:** The following administrative details and a protocol content summary should follow the title page:

- Contents page lists of relevant sections and sub-sections with corresponding page number.

- Signature page is signed by senior members of the research team and dated to confirm that the version concerned has been approved by them.

- Contact details for the research team members include postal, e-mail addresses and telephone numbers.

(iii) Project summary: The summary should be distinctive, concise and should sum up all the essentials of the protocol.

(iv) Background: The background to the project should be concise and refer to the subject straight forwardly. In writing the review, attention should be drawn to the positives, negatives and limitations of the studies quoted. Introduction is concluded by explaining how the present study will benefit the community. The literature review should logically lead to the statement of the aims of the proposed project and end with the aims and objectives of the study. The review should include the most recent publications in the field and the topic of the research should be selected only after completing the literature review and finding some gaps in it.

(v) Introduction should briefly answer the importance of the topic, the gaps/lacunae in the literature, the purpose of the study and benefits for society, from the study.

(vi) Study objectives: The aims should be explicitly stated. These should be confined to the intention of the project and they should arise from the literature review.

The study aims or objectives emerge from the study questions/hypotheses. They are answers to what are the possible responses to the research question or hypothesis under analysis and measure. Aims should be logical and coherent, feasible, concise, realistic, considering local conditions, phrased to clearly meet the purpose of the study and related to what the specific research is intended to accomplish. For example, to evaluate the knowledge level regarding dental caries in primary school children in KSA (this is not detailed). The following should be added: causes, treatment, preventive measures, etc.

The objectives should be (SMART objective): Specific, Measurable, Achievable, Relevant and Time based.

Specific Aims: Details of each objective that will finally lead to the achievement of this goal should be stated. Specific aims, one by one should be listed concisely. It is good practice not to include too many aims in the study (2–5 best); too many objectives often lead to inaccurate and poorly defined results. Furthermore, aims should be achievable, realistic and specific with no general and ambiguous statements. They should be stated in action verbs that illustrate their purpose: i.e., "to determine, to compare, to verify, to calculate, to reduce, to describe, etc."

Secondary objectives (optional): These are referred to as ancillary and minor objectives that could be studied during the course of the study. The formulation of objectives helps focus the study and to avoid the collection of any unnecessary data and hence organize the study in clear and distinct stages

Hypothesis: This is a statement based on sound scientific theory that recognizes the predicted correlation between two or additional assessable variables. It is always developed in response to the purpose statement or to answer the research questions posed. Furthermore, hypothesis transforms research questions into a format amendable to testing or into a statement that predicts an expected outcome.

Types of hypothesis statements:

Null hypothesis: A null hypothesis is the statement that there is no actual relationship between variables (H_0 or HN). It may be read as there is no difference between the groups to be compared and no relationship between the exposure and outcome under investigation. H_0 states the contradictory to what the researchers expect.

The conclusion of the investigators will either keep a null hypothesis or reject it in support of an alternative hypothesis. This does not essentially mean that H_0 is accurate when not rejecting it as there might not be an adequate proof against it.

Alternative hypothesis: An alternative hypothesis is a statement that suggests a potential outcome that the researcher may expect (H1 or HA). This hypothesis is derived from previous studies where an evident difference between the groups to be compared is present. This is recognized only when a null hypothesis is rejected. Practically, hypotheses are stated in the null form because they have their inferential statistics. Such hypotheses of no difference will be challenged by researchers and the result of the statistical testing gives the probability that the hypothesis of no difference is true or false

(vii) **Methods and Materials:** It should describe in detail the 'Where', 'Who', 'How' the research will be conducted. It explains the study design and procedures and techniques used to achieve the proposed objectives. It defines the variables and demonstrates in detail how the variables will be measured. It details the proposed methodology for data gathering and processing.

Methodology is an important part of this protocol. This assures that the hypothesis will be confirmed or rejected. It also refers to a thorough strategy to attain the objectives. The methods and materials are divided into various subheadings:

a) **Study design (cross-sectional, case-control, intervention study, RCT, etc.):** A proper explanation should be given as to why a particular design was chosen (on the basis of proposed objectives and availability of resources). A study design is in fact the researcher's general plan to acquire the answer(s) to the hypothesis being tested. Here, strategies will be applied to develop balanced, correct, objective and meaningful information. They explained the methods that will be used to collect and analyse data. Proper selection of the study design is important to attain reliable and valid scientific results. Ethics, logistic concerns, economic features and scientific thoroughness will determine the design of the study. Here, a chief concern is given to the legality of the results, including potential bias mystifying issues. Randomized controlled clinical trial is the best to document a causal relationship between an exposure and its outcome

b) **Study population:** It describes in detail about the study subjects, all aspects of the selection procedure and sample size calculation. Proper definition of eligibility, inclusion, exclusion and discontinuation criteria of the study subjects should be stated. Allocation of subjects to study arms should be explained and described in details bearing in mind the concealment and randomization process.

c) **Sample size:** Sample size calculation is recommended for economical and ethical reasons. The calculation of the sample size must be explained including the power of the sample. The sampling technique should be mentioned, e.g., randomization that will be used in order to obtain a representative sample for the target population. Each step involved in the recruitment of the study subjects should be described according to the selection criteria (inclusion and exclusion criteria).

"Informed consent" should be mentioned (Permission granted in full knowledge of the possible consequences).

d) **Proposed intervention:** Full description of the proposed intervention should be given. Here, all the activities and actions should be recorded and thoroughly explained in their order of occurrence.

- When using drugs, both scientific and brand names should be mentioned followed by the name of the manufacturing company, city, and country. The drug route, dosage, frequency of administration, and the total duration of treatment with the drug should be mentioned.

- When using an apparatus, its name should be given followed by the name of the manufacturer, city and country.

The involved personnel should precisely define:

- Who will be responsible for the interventions?

- What activities will each personnel perform and with what frequency and intensity?

e) Data collection methods, instruments used

Data collection tools are

- Retrospective data (medical records)

- Questionnaires

- Interviews (Structured, Semi-Structured)

- Laboratory test (literature or personal knowledge should be referenced, if established test, or description should be provided in details, if not established)

- Clinical examinations

- Description of instruments, tools used for data collection, as well as the methods used to test the validity and reliability of the instrument should be provided.

f) Data Management and Analysis Plan: This section should be written following statistical advice from a statistician. The analysis plan and the statistical tests will be used to check the significance of the research question/hypothesis with appropriate references described. Names of variables that will be used in the analyses and name of statistical analysis that will be performed to assess the outcome should be listed. If computer programs are to be applied, it is important to mention the software used and its version.

(viii) Project Management: Work plan-A work plan is an outline of activities of all phases of the research to be carried out according to an anticipated time schedule.

Proper time table for accomplishing each major step of the study should be defined. Assigning a time frame to each step in the trial will help organize the structure of the research trial. The personnel (investigators, assistants, laboratory technicians etc.) involved in the study or data collection, should be properly trained.

(ix) Strengths and Limitations: It is important to mention the strengths or limitations of the study, i.e. What study can achieve or cannot achieve important, to prevent wasteful allocation of resources.

(x) Ethical Considerations (Issues for Ethical Review and Approvals): It should indicate whether the procedures to be followed are in accord with the Declaration of Helsinki. In any case, the study should not begin unless approval from the ethics committee is received.

The following points should be explained:

- The benefits and risks for the subjects . The physical, social and psychological implications of the research.

- Details of the information to be given to the study patients, including alternative treatments/approaches.

- Information should be provided on the free informed consent of the participants. Information form should contain: Justification for research, outline of study, risks, confidentiality, and voluntary participation should be told patients about the freedom to withdraw from the study whenever they wish to. Confidentiality indicates how the personal information obtained from the patient will be kept secret (Data safety).

(i) Operational Planning and Budgeting (Budget Summary): Outline the budget requirements showing head wise expenditure for the study-manpower, transportation, instruments, laboratory tests, and cost of the drug. A budget estimate is to be attached to the annexure. All costs including personnel, consumables, equipment, supplies, communication, and funds for patients and data processing are all included in the budget. Each item should be justified.

(ii) Reference System: Referencing is the regular method for recognizing information taken from other researchers' work. A proper citation will enable the readers to follow-up any reference of interest. Plagiarism refers to claiming and acquiring someone else's ideas, an action that is considered a criminal action.

Failure to reference an idea which is found in the research or to acknowledge the work of other team members in a team assignment falls under the category of plagiarism. Therefore, referencing is an extremely important aspect of the research protocol. The two most commonly used citation systems in clinical writing are the Vancouver system and the Harvard system. The choice of referencing system depends upon the funding organization where the research protocol is being submitted. These frequently identify their preferred system of referencing and this should be strictly adhered to. The most common style used in the dental literature is the Vancouver style.

(iii) Annexure:

The following annexes are to be attached to the end of the protocol:

- Informed consent form.
- Letters from ethics committees.
- Study questionnaire (copies of any questionnaires or draft questionnaires).
- Case Record Forms (CRFs).
- Budget details.
- Curriculum Vitae (CV) of the chief investigator and co-investigator and their role in the study. This will ensure that the role of each investigator is well defined.

The protocol should adequately answer the research question. The research design must be sound enough to yield the expected knowledge. It should provide enough detail (methodology) that can allow another investigator to perform the study and arrive at comparable conclusions. Here, the proposed number of participants is reasonably justified and the scientific design is adequately described. The most difficult stage of conducting a research project is the preparation of a protocol that results in a short yet comprehensive document that clearly summarizes the project. Such proposal is considered successful when it is clear, free of typographical errors, accurate and easy to read.

Biostatistics in Pharmaceutical Product Development

Statistical concepts and tools have been successfully applied for decades in such sectors as chemicals, automobile manufacturing, and computer chip manufacturing, but their use in the far more regulated pharmaceutical industry presents some unique challenges. For example, many pharmaceutical companies hesitate to invest heavily in large-scale manufacturing-process quality before a drug is approved for marketing because failure in the clinic means product failure. Because time from inception to clinical approval may span 12–15 years, 60%–75% of product patent life may have expired by the time the Phase III (confirmatory) trials have been completed. Even after a successful set of Phase IIA (dose determination) and Phase IIB (proof of concept) clinical trials are completed, 40%–50% of product patent life may have expired. Additionally, one in three drugs is expected to fail in Phase I (first time in man) trials for assessing safety, tolerability, and drug blood levels. Pharmaceutical companies are therefore under considerable economic pressure to file for approval of the new drug application (NDA) with regulatory authorities as soon as possible after completing successful Phase III clinical trials, when a considerable amount of money must be invested in product launch. As a result, quality and process understanding initiatives must compete for time and money with potential losses from considerable 'at-risk' development activities.

Nevertheless, a number of factors are converging to increase the need for sophisticated, statistics driven approaches to quality and process understanding in the pharmaceutical industry, including:

1. Regulatory Trends.
2. Inherent Characteristics of Pharmaceutical Manufacturing.
3. Economic Pressures
4. Increased Need for Effective Technology Transfer.

Data Presentation for FDA Submission

Currently FDA accepts electronic submission of data. The regulatory process of electronic submission can be divided into four steps:

1. Early planning and management of the project
2. Document preparation
3. Dossier publishing, and
4. Dossier submission

The FDA is responsible for protecting the public health by ensuring the safety, efficacy, and security of human and veterinary drugs, biological products, and medical devices; and by ensuring the safety of our nation's food supply, cosmetics, and products that emit radiation. FDA is responsible for advancing the public health by helping speed innovations that make medical products more effective, safer, and more affordable and by helping the public get the accurate, science-based information they need to use medical products and foods to maintain and improve their health.

As electronic submissions continued to make progress, the Clinical Data Interchange Standards Consortium (CDISC) emerged, addressing the data component of regulatory submission standards. CDISC is a global, open, multidisciplinary, non-profit organization that has established standards to support the acquisition, exchange, submission, and archive of clinical research data and metadata. The CDISC mission is to develop and support global, platform-independent data standards that enable information system interoperability to improve medical research and related areas of healthcare. The continued collaboration of ICH, the FDA, and CDISC has enabled the development of electronic submission standards supported by technical documentation and specifications that are constantly being refined, updated, and implemented.

The binding guidance are:

➢ The Agency may Refuse-To-File (RTF) for NDAs and BLAs, or Refuse-To-Receive (RTR) for ANDAs if an electronic submission does not submit study data in conformance with the required standards specified in the Catalog.

➢ After the publication of this guidance, all studies with a start date 24 months after the publication date must use the appropriate FDA-supported standards, formats, and terminologies specified in the Catalog for NDA, ANDA, and certain BLA submissions and 36 months for IND studies. (This means that most NDA/BLA and ANDA studies beginning after December 17, 2016 and December 17, 2017 are bound by this guidance, respectively.)

➢ If the application is not submitted electronically, the sponsor may receive an RTF or RTR.

A submission data package requires a significant amount of internal and external collaboration in order to identify appropriate data standards,

- To develop a solid rationale for the data standards approach,

- To gain internal consensus, and

- To implement any required conversions.

The FDA is able to provide valuable input into this discussion and should be used as a resource in this planning stage. Open communication with the FDA will add value throughout the regulatory review process.

Management of Clinical Studies

Clinical Data Management (CDM) is a critical phase in clinical research. This can result in generation of high-quality, reliable, and statistically sound data from clinical trials. This helps to produce a drastic reduction in time from drug development to marketing. Team members of CDM remain actively involved in all stages of clinical trial right from beginning to completion. They should have adequate process knowledge that helps maintaining the quality standards of CDM processes. Various procedures in CDM include

- Case Report Form (CRF) designing,
- CRF annotation,
- Database designing,
- Data-entry,
- Data validation,
- Discrepancy management,
- Medical coding,
- Data extraction, and
- Database locking

These are assessed for quality at regular intervals during a trial. In the present scenario, there is an increased demand to improve the CDM standards to meet the regulatory requirements and stay ahead of the competition with the help of faster commercialization of product. With the implementation of regulatory compliant data management tools, CDM team can meet these demands. Additionally, it is becoming mandatory for companies to submit the data electronically. CDM professionals should meet appropriate expectations and set standards for data quality and also have a drive to adapt to the rapidly changing technology. This highlights the processes involved and provides the reader an overview of the tools and standards adopted as well as the roles and responsibilities in CDM.

CDM is the process of collection, cleaning, and management of subject data in compliance with regulatory standards. The primary objective of CDM processes is to provide high-quality data by keeping the number of errors and missing data as low as possible and gather maximum data for analysis. To meet this objective, best practices are adopted to ensure that data are complete, reliable, and processed correctly. This has been facilitated by the use of software applications that maintain an audit trail and provide easy identification and resolution of data discrepancies. Sophisticated innovations have enabled CDM to handle large trials and ensure the data quality even in complex trials.

CDM has guidelines and standards that must be followed. Since the pharmaceutical industry relies on the electronically captured data for the evaluation of medicines, there is a need to follow good practices in CDM and maintain the standards in electronic data capture. These electronic records have to comply with a Code of Federal Regulations (CFR), 21 CFR Part 11. This regulation is applicable to records in electronic format that are created, modified, maintained, archived, retrieved, or transmitted. This demands the use of validated systems to ensure accuracy, reliability, and consistency of data with the use of secure, computer-generated, time-stamped audit trails to independently record the date and time of operator entries and actions that create, modify, or delete electronic records. Adequate procedures and controls should be put in place to ensure the integrity, authenticity, and confidentiality of data. If data have to be submitted to regulatory authorities, it should be entered and processed in 21 CFR part 11-compliant systems. Most of the CDM systems available are like this and pharmaceutical companies as well as contract research organizations ensure this compliance.

Society for Clinical Data Management (SCDM) publishes the Good Clinical Data Management Practices (GCDMP) guidelines, a document providing the standards of good practice within CDM. GCDMP was initially published in September 2000 and has undergone several revisions thereafter. The July 2009 version is the currently followed GCDMP document. GCDMP provides guidance on the accepted practices in CDM that are consistent with regulatory practices. Addressed in 20 chapters, it covers the CDM process by highlighting the minimum standards and best practices.

Clinical Data Interchange Standards Consortium (CDISC), a multidisciplinary non-profit organization, has developed standards to support acquisition, exchange, submission, and archival of clinical research data and metadata. Metadata is the data of the data entered. This includes data about the individual who made the entry or a change in the clinical data, the date and time of entry/change and details of the changes that have been made. Among the standards, two important ones are the Study Data Tabulation Model Implementation Guide for Human Clinical Trials (SDTMIG) and the Clinical Data Acquisition Standards Harmonization (CDASH) standards, available free of cost from the CDISC website (www.cdisc.org). The SDTMIG standard describes the details of model and standard terminologies for the data and serves as a guide to the organization. CDASH v 1.1 defines the basic standards for the collection of data in a clinical trial and enlists the basic data information needed from a clinical, regulatory, and scientific perspective.

Drug Development, Manufacturing Processes, and Analytical Methods

Following the discovery of an API, pharmaceutical development takes place along five parallel paths (Figure 3.4):

1. Clinical trials,
2. Preclinical assessment,
3. API development,
4. Drug product development (final dosage form), and
5. Analytical method development.

The objective of this work is the submission and approval of an NDA. Once clinical development begins, it usually drives the time lines for the other four development paths.

API Development: R&D, Tech Transfer, and Manufacturing

The API in a medication or vaccine is the substance or organism that fights the symptom or illness that is being treated or induces immunity to the pathogen. After the discovery of a molecule that has pharmaceutical activity, work is done on this candidate API to further refine the material, to create the manufacturing process, and to scale it up to the desired production levels for full-scale manufacturing.

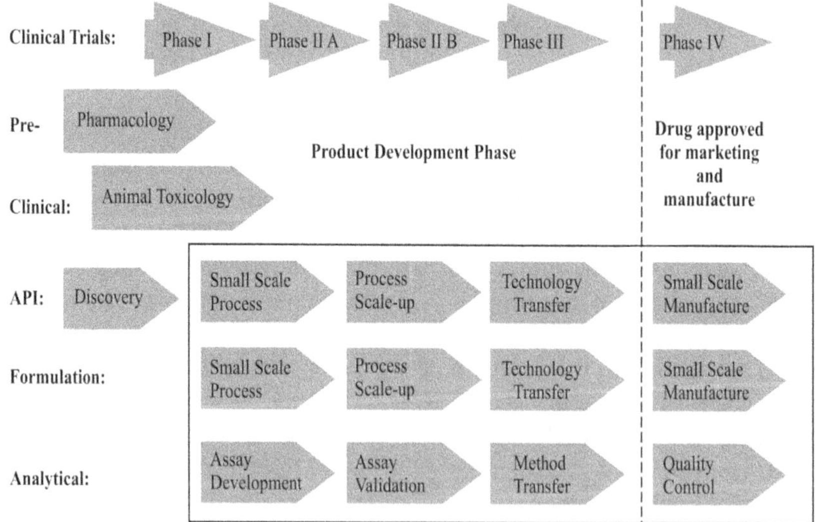

Fig. 3.6 Schematic of pharmaceutical development and manufacturing processes

Next, work is done to verify what was made and how to make it at a larger scale—for example, a move from 2–3-gm to 50-gm batches, which are needed to supply preclinical studies in animals. Work is then done to move to bigger batches (1–5 kg) using small-scale commercial equipment or scalable equipment (not beakers). With Phase III clinical studies, API manufacturing eventually grows to its largest scale to date, at least one-tenth of the commercial batch size. Much of the statistical thinking and many methods and tools used in the development and manufacture of an API are the same as those used for chemical processes. Also, much of the process and product development involves experimentation—the key tool, of course, being the design of experiments (DoE). Experimentation includes screening experiments, product/process understanding studies, regression modelling, process optimization, and robustness studies. In manufacturing, statistical process control (SPC) is used extensively to monitor and improve processes. On the improvement side, approaches using statistical techniques such as Six Sigma, Lean Manufacturing, PAT, Design for Six Sigma, and QbD are increasingly being used. Note that the preceding discussion of API applies to drugs developed from synthetic chemicals, the so-called "small molecules." Biologics and vaccines that are produced in biological systems are referred to as "large-molecule" drugs (e.g., proteins), as they are typically much larger molecules. Large and small-molecule drug production methods have a lot in common, with the principal difference being that large molecules are associated with many raw materials, numerous upstream and downstream process steps, various operating conditions, and numerous types of equipment. Because large-molecule drugs are produced from living organisms, variability is also higher and viral contamination can be an issue. Nevertheless, the associated statistical methods and approaches used are similar to those used for small-molecule drug production. Additional discussion of drug products from small and large molecules is included in this section.

Impact of Statistical Software and Information Technology

For drug-discovery research, the pharmaceutical industry has already witnessed a rapid increase in software for number crunching and information storage and retrieval. Part of this has been due to the genomic revolution and the associated 'OMIC' technological platforms for generating such data. However, with the advent of the process analytical technology (PAT) initiative, technology will again drive the need for more sophisticated software and information technology to store data, retrieve it and manage it. Furthermore, emerging statistical methods, as outlined in the previous section, will also require, to some degree, additional computing support or application development. Meanwhile, increased computing speed, storage, and interactive capabilities are also affecting quality and industrial statistics in the pharmaceutical industry. On one hand, the need will, in some cases, drive software creation. However, software availability will guide what statisticians and quality engineers are willing to do with statistics and graphics.

The increased availability of easy-to-use statistical commercial software allows scientists, engineers, and various technicians to perform their own analyses. Such software is useful not only for statisticians but for teaching and the enablement of (non-statistical) engineers and scientists.

A. Multiple Choice Questions

1. A regulatory affair (RA) is a profession, which acts as the interface between
   a) Pharmaceutical industry and drug regulatory authorities
   b) Pharmaceutical industry and the public
   c) Public and drug regulatory authorities
   d) Patients and pharmaceutical industry

2. The responsibility of regulatory affairs professional is to
   a) Remain updated with the company's yearly turnover
   b) Remain updated with the company's monthly turnover
   c) Remain updated with the company's product range
   d) Supply the physicians with accurate and complete information about the quality, safety and effectiveness of the product

3. Which of the following statements is correct?
   a) Estimation of safety and tolerability is conducted in a Phase II clinical study
   b) Effectiveness and short-term side effects of a drug-product are estimated in a Phase II clinical study
   c) Confirmation of therapeutic benefits was conducted in a Phase II clinical study
   d) None of the above

4. Which of the following statements is correct?

    a) The term '*in silico*' is used to define experimentation performed in the laboratory.

    b) The *in silico* studies are the slowest method of execution.

    c) The cost of *in silico* studies is high and have the ability to enhance the use of animals.

    d) The term '*in silico*' is used to define experimentation performed by computers.

5. Previously selected molecules of a drug are developed on the basis of

    a) *In-vitro* method   b) *Ex-vivo* method

    c) *In-silico* method   d) None of the above

6. The good laboratory practice (GLP) can be defined as

    a) A quality program related to organizational processes

    b) The conditions where non-clinical health and environmental safety are planned

    c) The conditions are performed, monitored, recorded, reported and archived.

    d) A quality program related to organizational processes and conditions where non-clinical health and environmental safety are planned, performed, monitored, recorded, reported and archived.

7. The toxic effects of substances leading to the discontinuation of their development are associated with

    a) The formation of toxic/reactive metabolites

    b) Associated with prolonged systemic drug exposure, the formation of toxic/reactive metabolites and/or possible drug-drug interactions.

    c) Prolonged systemic drug exposure

    d) Possible drug-drug interactions

8. The Regulatory Affairs department internal coordinates with other departments such as

    a) Drug development   b) Maintenance

    c) House-keeping   d) Electric

9. The Organization for Economic Co-operation and Development established a group of experts to formulate the first GLP principles with an attempt to:

   a) Avoid non-tariff barriers to the marketing of chemicals.

   b) Promote mutual acceptance (among member countries) of non-clinical safety data

   c) Eliminate the unnecessary duplication of experiments

   d) All of the above

10. Good Laboratory Practices are based on what?

    a) Manufacturing facility   b) Managing Director

    c) Study Director   d) Quality Control unit

11. The pharmacokinetic profile of a drug refers to which one out of the following?

    a) Area under the curve

    b) Concentration vs. time relationship

    c) C_{max} and t_{max}

    d) Clearance

12. A pre - IND meeting should be arranged with the FDA to discuss which of the following issue?

    a) The design of animal research, which is required to support the clinical studies

    b) Marketing strategy for the investigational drug or its product

    c) All of the above

    d) None of the above

13. The IND application must contain which one of the following information?

    a) Authorization letter from the Head of institution where the analytical testing is proposed to be done

    b) Certificate of the Head of Production unit where manufacturing of the drug or its product is proposed to be done

    c) Clinical Protocols and Investigator Information

    d) List of animals on which clinical studies have been conducted

14. Under which circumstances a new drug application (NDA) can be filed?

    a) When the drug/ drug-product is to be sold in USA

b) When the drug/ drug-product is to be sold in USA

c) When the drug/ drug-product is to be sold in USA

d) When the drug/ drug-product is found effective only

15. A copy of the application to conduct clinical trials in India should be submitted to whom?

a) The ethical committee

b) Local Municipal commissioner

c) Health Minister

d) Chief Minister of the concerned state

16. Which of the following is not a component of clinical research protocols?

a) Project summary

b) Title of the study

c) Preliminary studies

d) Analytical test reports

17. Which one of the following is not involved in the development of pharmaceuticals after the discovery of API?

a) Intermediate development of drug products

b) Final dosage form development

c) Analytical method development

d) Preclinical assessment

18. Which of the following is not to be attached to a clinical research protocol?

a) Budget details

b) Curriculum vitae of the principal investigator and co-investigator

c) Case record forms (CRFs)

d) Curriculum vitae of the Head of the organization.

19. Which is the odd one out of the following purpose of a clinical research proposal?

a) Clarification of the ethical considerations

b) Time and budget estimates

c) Formulation of hypothesis and objectives

d) Raising the question to be researched and clarification of its importance

20. Which one is not considered the content of an Investigator's Brochure?

   a) Title page

   b) Confidentiality Statements

   c) CV of the Investigator

   d) Results from non-clinical studies

B. Short Questions

1. What are the roles of regulatory department?

2. What are the documents prepared by the regulatory professionals?

3. What are *in-vitro* and *ex-vivo* assays?

4. What do you understand by the term's reproducibility, quality and reliability?

5. What do you understand of the desirable DMPK properties of new drug substances intended for oral administration?

6. What is meant by Investigational New Drug Applications (INDA)?

7. What are clinical research and bioequivalence studies?

8. What is the reference substance and test products as per the guidelines for bioequivalence study for orally administered drug substance?

9. What are the acceptance criteria for the assessment of bioequivalence?

10. Name the components of a clinical research protocol.

C. Long Questions

1. Justify that 'Drug Regulatory Affairs Department is the backbone of Pharmaceutical Industry.'

2. Write down the responsibilities of the regulatory affairs professionals.

3. Explain the term New Drug Applications (NDA).

4. Describe in brief the concept in Good Laboratory Practices.

5. Explain the recommended non-clinical assays of ADME.

6. What are the contents of the Investigator's brochure?

7. Explain briefly the types of hypothetical statements.

8. What is meant by biostatistics in pharmaceutical product development?

9. What is meant by the term, protocol? What are the purposes of clinical research protocol? How can this protocol be reviewed?

10. How the results of a clinical study be reported?

MCQs Answers

1	2	3	4	5	6	7	8	9	10	11	12	13	14	15	16	17	18	19	20
a	c	b	d	c	d	b	a	d	c	b	a	c	b	a	d	a	d	b	c

CHAPTER 4

Quality Management Systems

- Introduction
- History of the Quality Movement,
- Whats is the Quality?
- Why is Quality Important?
- Quality Management System,
- Good Manufacturing Practices (GMP)
- Good Clinical Practice (GCP)
- International Conference on Harmonization (ICH)
- Validation
- Total Quality Management (TQM),
- Six sigma
- Standardized systems/ISO standards
- ISO 9000 standards ISO 14000,
- National Accreditation Board for Testing Laboratory (NABL)
- Good Laboratory Practices (GLB)

Introduction

The quality management system is for unified standards for evaluating the processes, without a system in place to establish procedure, monitor progress and evaluate performance, it is nearly impossible to consistently deliver a quality product to your customer. By using quality management systems, problems are identified and corrected as they arise, allowing you to be proactive and minimize the likelihood of reoccurrence. A QMS helps coordinate and direct an organization's activities to meet customer and regulatory requirements and improve its effectiveness and efficiency on a continuous basis. A quality management system (QMS) is defined as a formalized system that documents processes, procedures, and responsibilities for achieving quality policies and objectives.

History of the Quality Movement

As early as the 1950s, Japanese companies started quality movements throughout their organizations with the help of an American, W. Edwards Deming. His methods include statistical process control (SPC) and problem-solving techniques that were very effective in changing the mentality of organizations needing to produce high-quality products and services.

Deming believed that 85 percent of all quality problems were the fault of the management. To improve, management has to take the lead and put the necessary resources and systems in place. For example, consistent quality in incoming materials could not be expected when buyers are not given the necessary tools to understand the quality requirements of those products and services. Buyers should fully understand the quality requirements, how to assess the quality of all incoming products and services, as well as be able to communicate these requirements to vendors. In some well-managed quality systems, buyers are allowed to work closely with vendors and help them meet or exceed the required quality requirements.

Another individual in the development of quality control was Joseph M. Juran, who, like Deming, worked in Japanese organizations focusing on improving quality and gained a name for him. Juran also established the Juran Institute in 1979; its goals and objectives of help organizations improve the quality of their products and services. Juran defined quality as "fitness for use," meaning that the users should be able to rely any time on that product or service upto100 percent without any worrying about defects. If this is true, the product could be classified as fit for use.

Quality of design could be described as what distinguishes a Yugo from a Mercedes-Benz and involves the design concept and specifications. The quality of a product or service is only as good as its design and intention. Thus, it is important to include quality issues in the design process, as well as to have in mind during the design phase, the difficulties one might have in replicating the product or service with the intended quality level.

What is the Quality?

According to the American Society for Quality, "quality" can be defined in the following ways:

❖ Based on the customer's perceptions about a product/service's design and how well the design matches the original specifications.

❖ The ability of a product/service to satisfy stated or implied applications.

❖ Achieved by conforming to established requirements within an organization.

Why is Quality Important?

The success of a business depends on the production of a product or service of quality higher than that produced by a competitor at a competitive price. When quality is the key to a company's success, a quality management system allows the organization to

➢ To keep up with and meet current quality levels,

➢ To meet the consumer's requirement for quality,

➢ To retain employees through competitive compensation programs, and

➢ To keep up with the latest technology.

Quality Management System

A quality management system is a management technique used to communicate with its employees about what is required to achieve the desired quality of products and services and what they should do to meet the specifications.

OR

A quality management system (QMS) is a set of policies, processes, and procedures. These are used to plan and execute production/development/service in the core business area of an organization to ensure customer satisfaction. ISO 9001 is an example of a Quality Management System.

OR

A quality management system (QMS) is a formalized system that documents processes, procedures, and responsibilities to achieve quality policies and objectives. A QMS helps to coordinate and direct an organization's activities to meet customer and regulatory requirements and to improve its effectiveness and efficiency continuously.

ISO 9001:2015, an international standard that specifies the requirements for quality management systems, is the most prominent approach to quality management systems.

While some use the term *QMS* to describe the ISO 9001 standard or the group of documents detailing the QMS, it refers to the entirety of the system. The documents only serve to describe the system. Quality management systems serve many purposes, including

- Improvement of processes
- Reduction of waste
- Lowering of costs
- Facilitating and identifying training opportunities
- Engaging staff
- Setting organization-wide directions
- Establishes a vision for the employees.
- Sets standards for employees.
- Builds motivation within the company.
- Sets goals for employees.
- Help fight against the resistance and to bring changes within organizations.
- Help direct the corporate culture.

Principles of a Quality Management System

- **Customer focus**

An organization always depends on its customers; therefore, it must understand and anticipate their current and future needs. Thus, it must work to satisfy their requirements and try to exceed their expectations.

- **Leadership**

Leaders are responsible for motivating and establishing the need of QMS and the policies of an organization. The leadership must strive to create and maintain an internal environment that fosters the total involvement of the staff in achieving the goals of the organization.

- **Involvement of the staff**

The staff members (employee) are the core resource of an organization. Only through their commitment to the organization and complete involvement, the skills of the workforce can be improved to the maximum level to meet their goals.

- **Process approach**

Any desired result can be achieved more efficiently if the necessary activities and resources are managed as a process.

- **System-centric management**

Identifying, understanding, and managing interrelated processes as a system can improve the efficiency of the performance of an organization in meeting its goals.

- **Continuous improvement**

The permanent objective of an organization should be to improve its global performance and this can only be possible through continuous improvement of its performance.

- **Evidence-based decision making**

Efficient decisions are based on the analysis of data and information.

- **Mutually beneficial relationships with the suppliers**

An organization and its suppliers are mutually interdependent. Therefore, a beneficial relationship between both parties can increase their mutual capacity and quality for creating value.

Types of QMS

1. *General*

    ISO 9001 – Quality Management System Requirement

    ISO 14000 – Environmental Management System

2. *Industry and Product Specific*

    TL 9000 – telecommunications

    API-Q1 – petroleum, petrochemical, and natural gas

3. *Process Specific*

GLP (FDA, EPA...) – conduct of non-clinical laboratory studies

ISO 17025 – general requirements for the competence of testing *and calibration laboratories*

Elements and Requirements of a Quality Management System

There are several elements of a quality system, and each organization has a unique system. The most important elements of a quality system include participative management, quality system design, customers, purchasing, education and training, statistics, auditing, and technology.

- ### *Participative Management*

The entire quality process, once started, will be an ongoing dynamic part of the organization, just like any other department such as marketing or accounting. It will also need the continuous focus of management. The implementation and management of a successful quality system involve different aspects that must be addressed continuously.

- ### *Vision and Values*

The starting point for the management and leadership process is the formation of a well-defined vision and value statement. This statement is used to establish the importance of the quality system and build motivation for the changes that must take place, such as the organization's plan to exceed customer expectations, commitment to fulfill a defined level of customer satisfaction, or commitment to manufacture or supply products with zero defects. The exact form of the vision and values is not as important as the fact that it is articulated and known by everyone . This vision and value statement is the driving force thatcan mold the culture required throughout the organization to maintain quality. Thus, the people and processes can determine whether there will be a change in quality. Vision and value are very important statements to set agendas for all other processes used to manage the quality system.

- ### *Developing the Plan*

The plan for the quality system may be different for a different organization, but there are similar characteristics:

- ❖ There should be clear and measurable goals.
- ❖ There should be financial resources available for quality.
- ❖ The quality plan should be consistent with the organization's vision and values.

The plan for the quality system might also include a pilot that would involve setting up small quality projects within the organization. This will allow the management to understand how well the quality system is accepted, learn from mistakes and will have greater confidence in launching an organization-wide quality system. The plan should provide some flexibility for employee empowerment; because, it has been demonstrated that the most successful quality system allows employees at all levels to provide input.

- *Communication*

Change, movement toward higher quality refers to effective communication for making the company a leader in moving the organization forward. Communication is the vital link between management, employees, consumers, and stakeholders. This communication also brings about closeness among all individuals and helps move toward the successful completion of long-term quality goals. Communication systems must allow the employees to provide feedback and provide possible solutions to issues the company must face. Management should allow formal and informal ways, such as employee feedback slips and feedback roundtable meetings.

- *Rewards and Acknowledgement*

Rewards, compensation, and acknowlededgment for achievements in quality are very effective tools to motivate employees. The employees say at the end of the day exactly what the management is trying to accomplish. Rewards, compensation, and acknowledgment may also be seen in a form of communication (these are tangible methods) that senior management wishes to let the employees know that quality is important. This could come in the form of individual or team rewards. Rewards, compensation, and acknowledgment take many forms, and it is up to the management to ensure that the type of program agrees with the goals and objectives of the quality system and the goals and objectives of the organization. Organizations have to determine the best and most cost-effective reward, compensation, and acknowledgment programs that can be effective in meeting these criteria. These programs motivate the managers, who in turn motivate their employees to strive for predefined goals.

Although any quality management system should be created to address the organization's unique needs; there are some general elements that all systems have in common, including

1. Data management

2. Internal processes

3. Customer satisfaction with product quality
4. Improvement opportunities
5. Quality analysis

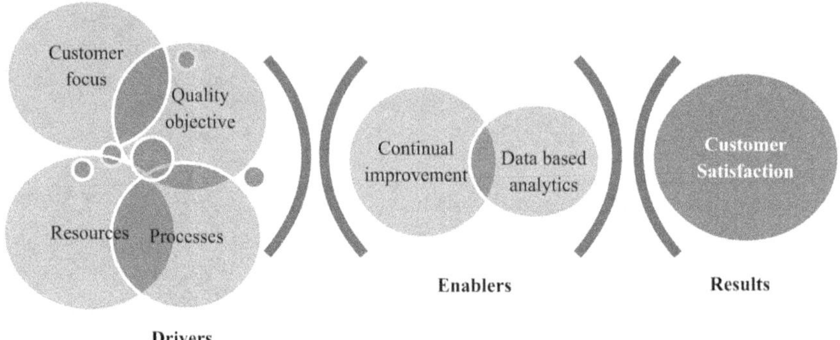

Fig. 4.1 Elements of a Quality Management System

Each element of a quality management system serves the purpose of meeting the customers' and organization's requirements. Each element of the QMS present must ensure the proper execution and function of the QMS.

Good Manufacturing Practices (GMP)

The concept of GMP was established when it was realized that quality is determined by the manufacturing process rather than by subsequent quality control. The term *Good Manufacturing Practice* was first used in 1962 in the Kefauver-Harris amendment to the Food and Drug Act in the United States; the FDA has set the pace for the global development of GMP. In the 1970s and 1980s, GMP became the subject of regulations in most countries. In 1969, the WHO published its GMP guidelines. The PIC was established in 1970 by the EFTA. Issued a GMP guide based on the WHO document. In Japan, GMP was established in 197 4 and enforced in 1975. The EU published GMP guidelines in January 1989, which served as the basis for the new PIC GMP guidelines of 1989 and the 1992 edition of the WHO GMP guidelines. In 1985, the Association of Southeast Asian Nations (ASEAN)-Brunei, Indonesia, Malaysia, the Philippines, Singapore, and Thailand-published the ASEAN GMP Guide. The present WHO GMP guidelines are based on the EEC and the ASEAN guidelines and are strongly influenced by the ISO 9000 series issued by the International Organisation for Standardisation (ISO).

GMP in Manufacturing Process

An increasing quantity of the drug material is needed as each stage progresses. The demand for material grows from milligrams to grams and kilo-grams. A development program is phased in to meet this demand by initially producing drugs on a laboratory scale. It then progresses to pilot plant scale to provide more drug material for clinical trials and finally implements procedures, processes, equipment, and setting up of a manufacturing plant for large-scale commercial production. Broadly, the drug development program covers the following.

- *Raw Materials*

Raw materials have a significant impact on the manufacturing process. Issues such as availability and reliability of supply, reactivity, toxicity, handling, and storage have to be considered. Cost is another factor to take into account. Often, trade-offs between costs, manufacturing processes, and yields are considered.

- *Safety*

Production of the requisite drug substance called the active pharmaceutical ingredient (API) or bulk pharmaceutical chemical (BPC), may involve materials, solvents, or volatile intermediates, toxic, or even explosive. The development program has to determine the appropriate manufacturing processes to ensure that safety is not compromised and the API can be produced and purified to remove impurities and toxic residues. The formulation of finished products with excipients/additives should ensure the safety of the formulated product.

- *Reproducibility of Manufacturing Processes*

GMP aims to ensure the manufacture of safe, potent, pure, and effective drugs consistently. The development program is used to evaluate procedures and processes that can be implemented in a large-scale manufacturing environment to ensure the drug product conforms to the intended safety, potency, purity, effectiveness, and consistency on a routine basis.

- *Environmental Factors*

In addition to conformance to GMP, the manufacturing plant has to comply with local environmental legislation. This may cover materials transportation, handling, storage, and disposal. The manufacturing plant is set up with systems for controlling gaseous emission, decontamination of solid waste, and treatment of liquid discharge. All these factors are evaluated in the development program.

- *Manufacturing*

The manufacturing of drugs, whether the API or finished dosage form, must comply with GMP regulations. Figure no shows the implementation of the GMP concepts in drug manufacture. The first requirement for GMP manufacture is the availability of trained personnel. Other requirements are

o Raw materials that conform to specifications Water, in the form of Purified Water or Water-for-Injection, as required, an environment with appropriate control for temperature, pressure, relative humidity. For the aseptic production, clean room conditions for monitoring and controlling for particles and bio-burden contamination are necessary.

o Approved specifications for the intermediates and finished drug product.

o Equipment validated and maintained with current calibration processes developed and validated to ensure the production of a pure and consistent product.

o Operating procedures are written down, detailing each manufacturing step.

o Approved batch records register all relevant information during the manufacturing process.

o Quality records document all tests of the raw materials, intermediates, and finished product.

o Validated cleaning procedure ensures that the reaction vessels and containers do not carry contaminants.

o Validation of sterilization or aseptic processes, as applicable.

o Satisfactory performance in the stability program.

GMP Inspection

Regulatory authorities inspect the GMP facilities to ensure that the drugs are manufactured according to the GMP requirements. inspection programs used in most developed countries are similar. The following is an example of the FDA inspection program. Drug Manufacturing Inspections Program, February 2002, states that the strategy for inspection as Evaluating through factory inspections, including the collection and analysis of associated samples, the conditions, and practices under which drugs and drug products are manufactured, packed, tested and held, and monitoring the quality of drugs and drug products through surveillance activities such as sampling and analyzing products in distribution.

FDA conduct inspections once every 2 years. The FDA has adopted a system approach to GMP Inspection. A GMP facility is divided into six systems:

- *Quality System:* This consists of procedures and specifications to assure the overall compliance of the facility. Quality control, change control, batch release, internal audits, quality records are part of the Quality System.

- *Facilities and Equipment System:* This includes (i) Buildings and facilities along with maintenance; (ii) Equipment IQ, OQ, calibration, maintenance, cleaning, and validation of cleaning processes; (iii) Utilities such as HVAC, compressed gases, steam, and water systems.

- *Materials System:* This is concerned with the segregation and storage of raw materials, components, and finished products, inventory control, and distribution of finished products.

- *Production System*: This includes manufacturing processes, sampling and testing, batch records, and process validation.

- *Packaging and Labeling System*: This includes control and issuance of labels, pack-aging operations, and validation of these operations.

- *Laboratory Control System*: It includes laboratory test methods, stability pro-gram, and analytical method validation.

FDA performs two types of inspections: (1) surveillance and compliance inspections. Surveillance inspections are biennial inspections. (2) Compliance inspections are to followup on non-compliance and previous corrective actions. Compliance inspections also include "Cause Inspections," which are inspections to audit a specific problem that has come to the FDA's attention, for example, product recall and the industry or public complaints.

There are two options of inspections: (i) Full Inspection Option, and (ii) Abbreviated Inspection Option:

- *Full Inspection* Option is surveillance or compliance inspection that is thorough and gives the FDA with a deep understanding of the CGMP program in a manufacturing facility. This type of inspection is conducted when the FDA has little knowledge about the facility, such as a new facility or where the facility has a history of non-compliance or when the FDA has doubt about the facility's quality system. The

Full Inspection Option audits at least four systems in the facility, one of which must be the quality system.

- *Abbreviated Inspection* Option is surveillance or compliance inspection. It is a shortened inspection. This is performed when the facility has a satisfactory CGMP compliance record and there are no product problems or there have been few changes since the last inspection. At least two systems are audited, including the quality system.

After inspection, the inspector prepares a detailed *Establishment Inspection Report (EIR).* This is the FDA's primary record for the inspection. Time is given to the manufacturer to respond to the deficiencies found and recorded in Form FDA-483. Failure to comply with a satisfactory resolution to the deficiencies found will result in the FDA sending out a Warning Letter notifying the manufacturer to comply. If the manufacturer is unable to resolve the deficiency after the deadline set by the FDA, FDA may proceed to prosecute the manufacturer with an injunction. The injunction is a court order called Consent Decree, and the manufacturer may be required to cease operations until the problem is rectified.

Some typical problems found in GMP inspections are

- *Out-of-Specifications:* Insufficient investigations to determine the root cause of problems and issues are not closed promptly.

- *Product Sterility:* The tests performed are superficial and not validated.

- *Environment Monitoring*: Personnel are not monitored, inadequate monitoring, microorganisms are not monitored, no identification of contaminants, alert limits for contaminants are set too high.

- *Raw Materials, Components, and Finished Product*: No or insufficient audit procedure, test methods lack validation.

- *Training:* There is no training plan, training is not documented, the investigation of problems in manufacturing does not lead to the retraining of staff .

- *Materials:* There was no segregation of materials.

- *Calibration:* Lack of scheduling and lapsed calibration validity.

- *Documentation:* Manufacturing steps are not signed off, changes are not explained, obsolete copies of the document are used.

- *Process:* Personnel not following set procedures, procedures are not validated

- *Internal Audit and Review:* Infrequent internal audit performed, superficial audits insufficient to reveal problems, no follow-up on issues observed at internal audits.

- *Management:* Lack or insufficient commitment to GMP issues.

Examples of some of the typical violations leading to drug product recall according to the FDA are

- cGMP deviations

- Sub-potent products

- The product failed the US Pharmacopoeia dissolution test requirements

- Product failed endotoxin/pyrogen tests

- Presence of contaminants in the products

- Label mix-up on the product

- The stability data do not support expiration date

- Product lacks stability

- Product failed content uniformity

- The product failed the pH test requirements.

Product Quality Review

It is the responsibility of the pharmaceutical company to maintain a robust GMP system while manufacturing the drug product. Regular periodic reviews of all licensed drug products to verify the consistency of existing processes, appropriateness of specifications for starting materials and finished products are to be conducted. These reviews would highlight trends and identify processes for improvement and rectification. The review should cover:

- Starting materials and primary product contact packaging materials

- Critical in-process controls and finished product results

- All the failed batches and their investigations

- Significant deviations or non-conformances. Their investigations and CAPA taken

- All changes were carried out to processes or analytical methods

- Manufacturing variations

- Stability monitoring programs and adverse trends

- Returns, complaints, and recalls

- Adequacy of the previous product process or equipment corrective actions

- Variations in marketing authorization and post-marketing commitments

- Qualification status of critical equipment and utilities.

Pharmaceutical companies have to ensure that GMP is under control at all times for the manufacture of drug products, and processes, conditions, and specifications follow the CMC information submitted to regulatory authorities. In reality, as time goes by, there are inevitable changes to processes, conditions, and specifications. The drivers for these changes are innovations, continual improvements, the results of process performance and product quality monitoring, and corrective and preventive actions. These post-market approval changes are considered in manufacturing variations. Manufacturing variations are notified or filed to the authorities immediately where prior approval is needed for major changes. Minor amendments of the lesser impact, are communicated to the authorities annually.

A change management system should be based upon sound principles, as presented in the ICH Q10 guideline, Pharmaceutical Quality System, which describes a comprehensive model for an effective pharmaceutical quality system. In the US, manufacturing variations are classified as major, moderate, or minor depending upon their degree or potential to adversely affect the drug product's identity, strength, purity, or potency.

The classifications of changes are determined on a risk-based approach:

- **Major Change**: A change that has a substantial potential to harm the safety or effectiveness of the product. Major changes require the submission of a Prior Approval Supplement (PAS) to the FDA, for which the FDA must approve before distributing the product made using the change.

- **Moderate Change**: A change that has a moderate potential to harm the safety or effectiveness of the product. Moderate changes require the submission of a changes being affected in the 30-Day Supplement to FDA at least 30 days before distributing the product made using the change.

- **Minor change**: A change that has a minimal potential to harm the safety or effectiveness of the product.

In Europe, the type of variation is divided as below:

Type IA Variation: Any well-defined minor change, to be notified within 12 months following implementation

- Type IA in Variation: Minor change; to be notified immediately after the changes have been implemented

- Type IB Variation: Minor change; this procedure is also considered a notification, but a regulatory authority assessment is made within 30 days; Type IB variation needs to be approved before implementation

Type II Variation: Any major change to the marketing authorization holder's proposed documentation; a Type II variation is a product change that does not meet Type IA and Type IB classifications or criteria, but is not so extensive as to require a line extension or new application procedure; Type II needs to be approved before implementation.

Line Extension Application: A major change that requires a full assessment.

GMP Regulatory

Regulatory guidelines are dynamic; they are revised and updated from time to time to implement new technologies, data, or information. Therefore, manufacturers have to keep abreast with regulatory developments by following current Good Manufacturing Practice (cGMP). GMP regulations came into effect in the United States in 1963. The implementation of GMP is the result of several tragedies in the United States and around the world. Some are mentioned here-

- In 1902, several children died after being administered contaminated diphtheria antitoxin.

- In 1937, 107 people died when the drug sulfanilamide was wrongly described.

- In 1955, 10 children died after being given improperly inactivated polio vaccine.

- In the 1960s, untold physical and emotional damage was caused by thalidomide.

So Legislation was introduced to ensure that drugs are safe for the patients and effective for treatment. The Food and Drug Administration (FDA) is responsible for ensuring that drug manufacturers comply with GMP regulations in the United States. GMP is defined by FDA as "federal regulation setting minimum quality requirements that drug, biologics, and device manufacturers must meet. Its intended result is total quality assurance and product control.

The US FDA GMP is coded by the following regulations:

- 21 CFR Part 210: Current Good Manufacturing Practice in Manufacturing, Processing, Packing, or Holding of Drugs;
- 21 CFR Part 211: Current Good Manufacturing Practice for Finished Pharmaceuticals
- 21 CFR Part 600: Biological Products:
- 21 CFR Part 610: General Biological Products Standards.

The areas where controls should be implemented are

- Organization and Personnel
- Building and Facilities
- Equipment
- Control of Components and Drug Product Containers and Closures
- Production and Process Controls
- Packaging and Labeling Controls
- Holding and Distribution
- Laboratory Control.

Note that the applicable regulations for small molecule drugs are 21 CFR Parts 210 and 211, and for biopharmaceuticals, the regulations are 21 CFR Parts 210, 211, 600, and 610. According to 21 CFR 210.1 (a), the regulations "contain the minimum current good manufacturing practice for methods to be used in and the facilities or controls to be used for, the manufacture, processing, packing, or holding of a drug to assure that such drug meets the requirement."

Good Clinical Practice (GCP)

Cases of violations of human rights and scientific misconduct or fraud in connection with clinical trials in the United States in the 1960s and 1970s led the FDA to issue regulations and to control compliance. Successful clinical trials for drug candidate lead to the filing of a marketing authorization application to the regulatory authorities. Once the marketing authorization is approved, a pharmaceutical firm is ready to manufacture the drug for commercial sale to the target patient group. The manufacture of the drug must be under Good Manufacturing Practice (GMP). GMP is a quality concept and consists of a set of policies and procedures for the manufacturers of drug products. These policies and procedures describe the facilities, equipment, methods, controls, and personnel training needed for producing drugs with the intended quality. The term Quality by Design (QbD) is used. Manufacturers are required to abide use GMP regulatory guidelines to ensure that drugs are pure, consistent, safe, and effective.

International Conference on Harmonization (ICH)

Today, harmonization is a major factor in the development of drug laws in the EU, Japan, and the United States. Based on previous harmonization efforts. The ICH began taking shape in 1989, when the Steering Committee was established. The tripartite harmonization of guidelines by the United States, Europe, and Japan with the formation of the International Conference on Harmonization (ICH). As noted, there are four major categories of harmonized guidelines prepared by ICH: Quality, Safety, Efficacy, and Multidisciplinary.

ICH guidelines are divided into four major categories-

A. **Quality guideline:** 11 topic headings – Stability, Analytical Validation, Impurities, Pharmacopeias, Quality of Biotechnological Products, Specifications, GMP, Pharmaceutical Development, Quality Risk Management, Pharmaceutical Quality System, Development and Manufacture of Drug Substances. A total of 47 documents.

B. **Safety guideline:** 10 topic headings – Carcinogenicity Studies, Genotoxicity Studies, Toxicokinetics and Pharmacokinetics, Toxicity Testing, Reproductive Toxicology, Biotechnological Products, Pharmacology Studies, Immunotoxicology Studies, Nonclinical Evaluation for Anticancer Pharmaceuticals, Photosafety Evaluations. A total of 15 documents.

C. **Efficacy guideline:** 9 topic headings – Clinical Safety, Clinical Study Reports, Dose-Response Studies, Ethnic Factors, GCP, Clinical Trials, Clinical Evaluation by Therapeutic Category, Clinical Evaluation, Pharmacogenomics: A total of 26 documents.

D. **Multidisciplinary guideline:** 8 topic headings

Quality Guidelines

Q1: Stability

- o Q1A Stability testing of new drug substances and products
- o Q1B Stability testing: photostability testing of new drug substances and products
- o Q1C Stability testing for new dosage forms

- o Q1D Bracketing and matrix designs for stability testing of new drug substances and products

- o Q1E evaluation of stability data

- o Q1F Stability data package for registration applications in climatic zones III and IV

Q2: Analytical Validation

- o Q2 Validation of analytical procedures: text and methodology

Q3: Impurities

- o Q3A Impurities in new drug substances

- o Q3B Impurities in new drug products

- o Q3C Impurities: a guideline for residual solvent

- o Q3D Impurities: a guideline for elemental impurities

Q4: Pharmacopoeias

- o Q4A Pharmacopeial harmonization

- o Q4B and Annexes Evaluation and recommendation of pharmacopeial tests for use in the ICH regions

Q5: Quality of Biotechnological Products

- o Q5A Viral safety evaluation of biotechnology products derived from celllines of human or animal origin

- o Q5B Analysis of the expression construct in cells used for the production of rDNA derived protein products

- o Q5C Stability testing of biotechnological/biological products

- o Q5D Derivation and characterization of cell substrates used for the production of biotechnological/biological products

- o *Q5E Comparability of biotechnological/biological products subject to changes in their* manufacturing process

Q6: Specifications

- o Q6A Specifications: Test procedures and acceptance criteria for new drug substances and new drug products; chemical substances

- o Q6B Specifications: Test procedures and acceptance criteria for biotechnological/biological products

Q7: Good manufacturing practice guide for active pharmaceutical ingredients

Q8: Pharmaceutical development

Q9: Quality Risk Management

Q10: Pharmaceutical quality systems

Q11: Development and manufacture of drug substances

Safety Guidelines

S1: Carcinogenicity Studies

- o S1 Rodent Carcinogenicity Studies for Human Pharmaceuticals
- o S1A Need for Carcinogenicity Studies of Pharmaceuticals
- o S1B Testing for Carcinogenicity in Pharmaceuticals
- o S1C(R2) Dose Selection for Carcinogenicity Studies of Pharmaceuticals

S2: Genotoxicity Studies

- o S2(R1) Guidance on Genotoxicity Testing and Data Interpretation for Pharmaceuticals Intended for Human Use

S3: Toxicokinetics and Pharmacokinetics

- o S3A Note for Guidance on Toxicokinetics: The Assessment of Systemic Exposure in Toxicity Studies
- o S3B Pharmacokinetics: Guidance for Repeated Dose Tissue Distribution Studies

S4: Toxicity Testing

- o S4 Duration of Chronic Toxicity Testing in Animals (Rodent and Nonrodent Toxicity Testing)

S5: Reproductive Toxicology

- o S5(R2) Detection of Toxicity Reproduce Medicinal Products and Toxicity to Male Fertility

S6: Biotechnological Products

- o S6 Preclinical Safety Evaluation of Biotechnology-Derived Pharmaceuticals

S7: Pharmacology Studies

- o S7A Safety Pharmacology Studies for Human Pharmaceuticals
- o S7B The Nonclinical Evaluation of the Potential for Delayed Ventricular Repolarization (QT Interval Prolongation) by Human Pharmaceuticals

S8: Immunotoxicity Studies for Human Pharmaceutical

S9: Nonclinical Evaluation of Anticancer Pharmaceuticals

S10: Photo safety Evaluation of Pharmaceuticals

Efficacy Guidelines

E1 The extent of population exposure to assess the clinical safety of drugs intended for long-term treatment of non-life-threatening conditions

E2A clinical safety data management: definitions and standards for expedited reporting

E2B (R3) Clinical safety data management: data elements for the transmission of individual case safety reports

E2C (R2) Periodic benefit-risk evaluation report

E2D Post-approval safety data management: definitions and standards for expedited reporting

E2E Pharmacovigilance planning

E2F Development safety update report Clinical study reports

E3 Structure and content of clinical study reports

E4 Dose-response information to support drug registration Ethnic Factors

E5 (R1) Ethnic factors in the acceptability of foreign clinical data Good clinical practice

E6 (R1) Good clinical practice clinical trials

E7 Studies in support of special populations: geriatrics

E8 General consideration of clinical trials

E9 Statistical principles for clinical trials

E10 Choice of the control group and related issues in clinical trials

E11 Clinical investigation of medicinal products in the pediatric population guidelines for clinical evaluation by therapeutic category

E12 Principles for the clinical evaluation of new antihypertensive drugs clinical evaluation

E14 Clinical evaluation of QT/QTc interval prolongation and proarrhythmic potential for non-antiarrhythmic drugs Pharmacogenomics

E15 Definition for genomic biomarkers, pharmacogenomics, genomic data, and sample coding categories

E16 Biomarkers related to drug or biotechnology product development: content, structure, and format of qualification submissions.

Multidisciplinary Guidelines

M1 – Medical Terminology (MedDRA)

M2 – Electronic Standards for Transmission of Regulatory Information (ESTRI)

M3 – Nonclinical Safety Studies

M4 – The Common Technical Document (CTD)

M5 – Data Elements and Standards for Drug Dictionaries

M6 – Gene Therapy

M7 – Genotoxic Impurities

M8 – Electronic Common Technical Document (eCTD).

Validation

In 2011, FDA updated its definition for Process Validation (Guidance for Industry- Process Validation: General Principles and Practices, January 2011) as the following: "The collection and evaluation of data, from the process design stage throughout production, which establishes scientific evidence that a process is capable of consistently delivering quality products."

The validation process is the documented evidence that provides a high degree of assurance to the desired result with predetermined compliance. The emphasis is the collection of data throughout the product life cycle of the drug and to ensure that quality is consistent. This concept of managing product life cycle quality is closely linked to ICH documents Q8, Q9, Q10, and Q11.

Types of validations
➢ *Analytical method validation*
➢ *Process validation*
➢ *Cleaning validation*
➢ *Equipment validation*

Analytical Method Validation: The purpose of analytical validation is to verify that the selected analytical procedure will give reliable results that are adequate for the intended purpose

Process Validation: Process validation also assures the repeatability of the process and decreases the risk of manufacturing problems, which increases the output of predetermined quality.

It can be of four types, which are as follows:

- Prospective validation
- Concurrent validation
- Retro-specific validation
- Revalidation.

There are three stages of process validation:

- **Stage 1-Process Design:** The commercial manufacturing process is defined as a stage based on the knowledge gained through development and scale-up activities.

- **Stage 2-Process Qualification:** During this stage, the process design is evaluated to determine whether the process is capable of reproducible commercial manufacturing.

- **Stage 3-Continued Process Verification:** ongoing assurance is gained during routine production that the process remains in a state of control.

A mindset of validated approach will help assure drug product quality, optimize the processes and reduce manufacturing costs. The approach to validation starts with the Validation policy and then the development of the Validation Master Plan (VMP), which details

- Validation policy
- Organization of validation activities
- Personnel responsibilities
- Facilities, systems, equipment, and processes to be validated
- Documentation structure and format
- Change control processes
- Planning and scheduling.

Equipment validation: Equipment validation is an established documented set up that proves that any equipment works correctly and leads to accepted and accurate results that are as follows:

- Design qualification
- Installation qualification
- Operational qualification
- Performance qualification

Scopes of DQ, IQ, OQ, and PQ: The following sets out the scopes generally expected to be performed for DQ, IQ, OQ, and PQ.

DQ:

- Verify
  - Design meets URS
  - The design complies with VMP
  - Software meets GAMP
  - All the utilities are available and validated
  - Support documents are specific and available
  - System/equipment/instrument can be calibrated
  - System/equipment/instrument can be maintained
  - There is training for operating staff
  - System/equipment/instrument is safe to both product and personnel
  - System/equipment/instrument conforms to applicable standards.

IQ:

- Verify
  - Purchase documents (e.g. purchase order, invoice)
  - The model number and serial numbers
  - Utilities connected as specified
- Check
  - Manufacturer and supplier
  - Physical damage
- Confirm the location and installation requirements per the recommendation of manufacturers
- Installation shall be conducted as per the instructions provided in the manual
- Ensure that all relevant documentation is received:
  - User manual
  - Maintenance manual
  - Calibration certificate of sensors/measuring instruments
  - Software manual
  - Parts list
  - Electrical drawings
  - Mechanical drawings.

OQ:

- Verify
  - o Alarm control and interlocks
  - o All the switches and push button function properly
- Perform the calibration requirements identified in the manual or established by the validation team
- Operate the equipment at different parameter settings, for example, low, medium, and high speed/time/temperature per operation manual to verify the operating control.
- Check software operations
- Establish procedures for operation, maintenance, and calibration
- Establish a training program for relevant staff.

PQ:

- Check the specific performance of system/equipment, for example, vial washing machine
- Perform visual inspection
- Check pH, conductivity, TOC of rinse water to WFI specifications
- Endotoxin
  - o Unwashed vials: less than 10 EU/mL rinse water
  - o Washed vials: less than 0.25 EU/mL rinse water
- Particulates
  - o Positive and negative controls
- Spiked with about 500 particles of 40 µm glass beads
  - o Rinse samples greater than 95% reduction
- Soiled samples
  - o Use of positive and negative controls
  - o Spiked with 10% sodium chloride or media/materials likely to encounter
  - o After washing, rinse samples should have no traces of soiled materials.

DQ is performed by the supplier of the equipment or system at the supplier's factory as part of a factory acceptance test (FAT). This ensures that all requirements of the equipment are recorded and tested in IQ, OQ, and PQ. IQ is based on the site acceptance test (SAT), OQ, and PQ are performed on-site at the GMP facility. For a GMP manufacturing facility, the validation activities are facility design, HVAC system, environment control, laboratory, and production equipment, water system, gases and utilities, cleaning, analytical methods, sterilization, and critical production processes. Validation protocols (IQ, OQ, and PQ) are prepared for each item, listing all critical steps and acceptance criteria. The deviations are reviewed and resolved before the validation activity proceeds to the next stage. After satisfactory execution of DQ, IQ, OQ, and PQ, process validation (PV) can be performed to verify that the process can be consistently and accurately replicated and that the outcome is a drug product of the requisite quality.

Total Quality Management (TQM)

TQM is a management approach in which quality is emphasized in every aspect of the business and organization. Its goals are aimed at the long-term development of quality products and services. TQM breaks down every process or activity and emphasizes that each contributes or detracts from the quality and productivity of an organization as a whole. Management's role in TQM is to develop a quality strategy that is flexible enough to be adapted to every department, aligned the organizational business objectives and based on the needs of customers and stakeholders. Once the strategy is defined, it must be the motivating force to be deployed and communicated for it to be effective at all levels of an organization. Some degree of employee empowerment is also encompassed in the TQM strategy and usually involves both departmental and cross-functional teams to develop strategies to solve quality problems and making suggestions for improvement.

Continuous Quality Improvement (CQI)

Continuous quality improvement came into existence in manufacturing as a different approach to quality and quality systems. It does not focus as much on creating a corporate quality culture, but more on the process of quality improvement by the deployment of teams or groups who are rewarded when goals and quality levels are reached. CQI allows individuals involved in the day-to-day operations to change and improve processes and work flows as they see fit.

CQI implementation attempts to develop a quality system that is never satisfied; it strives for constant innovation to improve work processes and systems by reducing time-consuming, low value-added activities. The time and resource saving can now be devoted to planning and coordination.

CQI has been adapted in several different industries. For example, in health care and other service sectors, it has taken on the acronym *FOCUS-PDCA* work:

Find a process to improve.

Organize to improve the process.

Clarify what is known.

Understand variation.

Select a process improvement.

Then move through the process improvement plan:

Plan- create a timeline, including all resources, activities, dates, and personnel training.

Do - implement the plan and collect data.

Check - analyze the results of the plan.

Act - act on what was learned and determine the next steps.

The *FOCUS-PDCA* acronym is an easy system for the management to communicate to teams, and it helps them stay organized and on track with the result in mind. The system has proven to be very successful for the CQI team approach.

Quality by Design

The pharmaceutical industry is alert for product quality, safety, and efficacy. Product quality has been increasing by implementing scientific tools such as QbD (Quality by Design). Scientific approaches will provide clear and sufficient knowledge from product development to manufacturing. These QbD tools will minimize the risk by increasing the output and quality. Nowadays, QbD approach has been successfully implemented in common formulation development. The USFDA has released specific QbD guidance for immediate and extended-release drug and biotechnological products. Regulatory authorities always propose the implementation of ICH quality guidelines Q8 to Q11.

According to ICH Q8 guidelines, QbD is defined as, "A systematic approach to development that begins with predefined objectives and emphasizes product, process understanding & process control, based on sound science and quality risk management." It means that design & development the formulation & manufacturing process to make sure predefined product quality. It requires an understanding of how product and process variables influence product quality. It is a systematic process of building quality into the final product. QbD requires the identification of all critical quality attributes and process parameters as well as determining the level to which any variation can impact the quality of the final product.

Concepts and Background of QbD

Quality by Design is a concept first outlined by Joseph M. Juran, in various publications. He supposed that quality could be planned. The concept of QbD was mentioned in ICH Q8 guidelines, which state, "To identify quality cannot be tested in products, i.e. Quality should be built into the product by design." In 1970, Toyota pioneered many QbD concepts to improve their early automobiles, since that time other industry technology, telecommunication and aeronautics have taken this concept and made QbD. In 1990, medical devices began to show that they incorporated many qualities by design aspects. In mid-2002 FDA published a concept paper on cGMP for the 21st century. These documents expressed a desire that companies build quality, safety, and efficacy into their new products as early as possible.

Objectives of QbD

- The main objectives of QbD are to ensure the quality of products, for that product and process characteristics are important to desired performance, must result from a combination of prior knowledge & new estimation during development.
- From this knowledge & data process measurement & desired attributes may be constructed.
- The experimental study would be viewed as positive performance testing of the model ability through Design space.
- Ensures a combination of product and process knowledge gained during development.

Benefits of QbD

For industry

- A better understanding of the process.
- Less Batch failure.
- Avoid regulatory compliance problems
- Ensures better design of products with fewer problems in manufacturing.

- Allows for continuous improvement in products and manufacturing process.

For FDA

- Enhances scientific base for analysis.

- Provide better consistency.

- Provides for more flexibility in decision-making.

- Ensures decisions are made on science and not on observed information.

Pharmaceutical Aspects: Traditional vs. QbD Approach

Advancement in pharmaceutical development and manufacturing by QbD can be explained against the traditional approach below (Table 4.1).

Table 4.1 Pharmaceutical Aspects: Traditional vs. QbD approach

Aspects	Traditional	QbD
Pharmaceutical Development	Empirical	Systematic, multivariate experiments
Mfg. process	Fixed	Adjustable within design space
Process control	Offline analysis wide or slow response	PAT is used for feedback and provides real-time
Product Specification	Based on batch data	Based on the desired product performance
Control strategy	Mainly by intermediate product and end-product testing	Risk-based, Controlled shifted upstream, real-time release
Lifecycle management	Post-approval changes needed	Continual improvement enables with in design space

1. In the QbD Process, first define Target Product Profile (TPP) and Quality Target Product Profile (QTPP), which describe the use, safety, and efficacy of the product.

2. Once TPP and QTPP have been identified, the next step is to identify the Critical Quality Attributes. (CQAs)

3. Perform the risk assessment that linking material attributes and process parameters that must be controlled to achieve the desired quality product.

4. After that, confirm the Design space (i.e. Requirement for in-process drug substance and drug product attributes) these conditions are established based on several sources of information.

5. Implement a control strategy for the entire process using in-process & end process controls.

6. Perform continuous improvement to get consistency in the quality of products (Fig. 4.2).

Elements of QBD

o ***Target Product Profile (TPP)-***Under this title, the target is an important word. The target is nothing but the result that we try to achieve. So, in this, we target the drug profile or target product that ensures desired quality, safety & efficacy. TPP is defined as, "A prospective summary of the quality characteristics of a drug product that ideally will be achieved to ensure the desired quality, taking into account safety & efficacy of the drug product." (ICH Q8) target product profile should include-

- Dosage form

- Route of administration

- Dosage strength

- Pharmacokinetics

- Stability

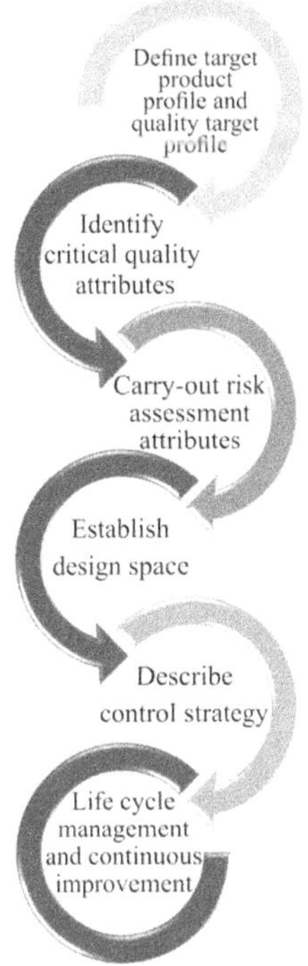

Fig. 4.2 Steps of Quality by Design

The TPP is a patient and labeling-centered concept because it identifies the desired performance characteristics of the product, related to the patient's need & it is organized according to the key section in the drug labeling. Pharmaceutical companies will use the desired labeling information to construct a target product profile. The TPP is then used to design the clinical trials, safety & ADME studies as well as to design the drug product, i.e. The QTPP.

o ***Quality Target Product Profile (QTPP)-*** QTPP is a quantitative substitute for aspects of scientific safety and efficacy that can be used to design and optimize a formulation and mfg. process. It should include quantitative targets for impurities, stability, and product-specific performance requirements. QTPP is not specification because it includes tests such as bioequivalence or stability that are not carried out in batch to batch release. QTPP should only include patient-relevant product performance. The Quality Target product profile is a term that is an ordinary addition of TPP for product quality. It guides formulation scientists to establish formulation strategies and keep formulations well-organized. QTPP is related to identity, assay, dosage form, purity, stability in the label.

o *Critical Quality Attributes (CQAs)*-A CQA has been defined as "a physical, chemical, biological, or microbiological property or characteristic that should be within an appropriate limit, range, or distribution to ensure the desired product quality." CQAs were identified through risk assessment as per the ICHQ9. Critical Quality Attributes are generally associated with the drug substance, excipients, intermediates, and drug product. Critical Quality attributes include the properties that impart the desired quality, safety, and efficacy. CQAs for biotechnological products are typically those aspects affecting product purity, stability. Drug product CQAs can be identified from the Target product profile. The use of strong risk estimation methods for the identification of CQAs is new to the QbD standard.

o *Risk Assessment*

Risk assessment is the linkages between material attributes & process parameters. It is performed during the lifecycle of the product to identify the critical material attributes & critical process parameters.

o *Control Strategy*

The control strategy is defined as, "A designed set of control, derived from current product and process understanding that assures process performance and product quality." A control strategy is designed to ensure that products of required quality will be produced consistently. Once a sufficient level of process understanding is achieved, a control strategy should be developed that assures that the process will remain in control within the normal variation in material attributes and process operating ranges. Control strategies may include input material controls, process controls, and monitoring, design spaces around individual or multiple unit operations, and final product specifications used to ensure consistent quality.

Elements of a Control Strategy
- Procedural controls
- In-process controls
- Batch release testing
- Process monitoring
- Characterization testing

- Comparability testing
- Constancy testing

The control strategy in the QbD standard is established via risk assessment that considers the criticality of the CQAs

o ***IFM (Impurity Fate Mapping)***: This is such an example in which raw material and process impurity sources are identified and their fate mapped throughout the process. Removing impurities is an essential element of the control strategy.

o ***Product Lifecycle Management and Continual Improvement***

Throughout the product lifecycle, companies have opportunities to evaluate modern approaches to improve product quality. After approval, CQAs would be monitored to ensure that the process is performed within the defined suitable variability that served as the basis for the field process design space. The primary benefit of an extended process design space would be a more flexible approach by regulatory agencies. Therefore, process improvements during the product life cycle are related to process consistency.

Application of QBD

QBD in Influenza Vaccines

Influenza (flu) is caused by influenza viruses and is primarily spread mainly by coughing, sneezing, and close contact with an infected person. Flu is a communicable disease that spreads around the US every winter in Oct.

Symptoms
- Fever/chills
- Sore throat
- Muscle aches
- Fatigue
- Cough
- Headache

Vaccination

Vaccination is the phenomenon of protective immunization. In the modern concept, vaccination involves the administration (injection or oral) of an antigen to obtain an antibody response that will protect the organism against future infections. Attenuated viruses are genetically modified pathogenic organisms that are made non-pathogenic &used as vaccines. Attenuated strains of some pathogenic organisms were prepared by prolonged cultivation for weeks, months, or years. Due to this, the infectious organism would lose its ability to cause disease but retain its capacity to act as an immunizing agent. The flu vaccine is the best protection against flu and its complications. The flu vaccine also helps prevent the spreading of flu from person to person. Flu vaccine cannot prevent all cases of flu, but it is the best protection against acute respiratory diseases.

Some people should not get this vaccine

1. If they have any severe, life-threatening allergies. e.g., If allergy to gelatin, antibiotics, or eggs, you may not get vaccinated.

2. If you are not feeling well, then do not get vaccinated (Table 4.2).

Table 4.2 Quality target product profile for vaccines

Parameters	Description
Mechanism of Action	Type A-VAX is a pentavalent vaccine containing the capsular polysaccharide of 5 serotypes, each linked to a recombinant non-infectious virus-like particle (VLP). This is expected to produce enhanced cellular antigen-specific protective immunity
Indication	Active immunization of 2-60-month-old infants and children for preventing disease-related illnesses due to causative agents
Primary Endpoints	70% reduction of confirmed disease within 1 year after dosing in the target population, safe and tolerable
Key Claims	Easy to administer, 0.5ml subcutaneous delivery in an outpatient setting using a 1ml syringe. Stability-2 years at RT
Formulation/Dosing	The sterile product, 3 doses, were administered 2 months apart. Label volume-0.5 ml Primary packaging: Single-dose vial, clear Type 1 glass. Secondary packaging: 10 vials/carton
Approvals and Recommendations	Expecting Advisory Committee on Immunization practices and other universal recommendations

Cell-Culture Based Influenza Vaccine Production

Objectives

1. Modern cell-culture technology potentially allows for quick, efficient production.

2. Production of cell-derived vaccines requires little-advanced planning & many respond to the event of the virus.

The influenza vaccine production process involves 5 fundamental steps:

1. Cell propagation/Preparation of substrate.
2. Virus propagation
3. Purification
4. Inaction and Splitting
5. Blending, Filling, & Approval.

1. ***Cell propagation/Preparation of substrate:*** Take the frozen, preserved cell culture from the WBC cell line and grow it in an incubator at 37°c. Then, these cells are first grown in a small volume of culture medium, due to which cells are grown and multiplied. They were then transferred into a successively larger container.

2. ***Virus propagation:*** Once a high number of cells have been produced, add influenza seed virus obtained from the WHO diagnostic laboratory into a cell containing bioreactor (Fermenter) where the virus infects the cells and multiplies and produces more virus particles. After several days, viruses destroyed the cell in bioreactors. The virus is harvested by removing the waste made by the cells and made non-infectious.

3. ***Purification:*** Using a centrifuge or chromatography, the virus is then separated from the cells and removed from the solution

4. ***Inactivation and splitting:*** A chemical process is used to inactivate the virus, stripping it of its ability to infect, for that formaldehyde is used, is called splitting. Then, the surface antigen is separated and extracted from the virus.

5. ***Blending, filling, and approval:*** Non-infectious solution is blended, concentrated, and filled into a sterile syringe.

By using QbD the following parameters should be controlled during the vaccine production process.

(i) *Cell propagation:* In this step, limiting the concentration of nutrients may be helpful for optimal cell growth. If they at a high nutrient concentration, then it inhibits cell growth. For that to do online monitoring of the concentration of the nutrient.

(ii) *Virus prorogation:* The following variable parameters are controlled during the fermentation process.

- pH: maximum effectiveness of fermentation can be achieved by continuous monitoring pH i.e. It required the most favorable pH.

- Temperature: Temperature control is important for a good fermentation process. If the temperature is lower than it causes reduced product formation & if it is higher than it affects the growth of organisms. To avoid this, bioreactors are equipped with a heating and cooling system as per the requirement to maintain the reaction vessel at an optimal temperature.

- Dissolved oxygen: Optimal supply of nutrients and oxygen, due to this it prevents the growth of toxic metabolic byproducts.

- Agitation: Good mixing also creates a favorable environment for growth and good product formation. If the agitation is excessive, then it damages the cells and increases the temperature of the medium.

- Foam formation: Avoiding this parameter antifoam chemicals are used such as mineral oils, vegetable oil, which lower the surface tension of the medium and cause foam bubble collapse. Also, mechanical foam control devices are fitted at the top of the fermenter.

(iii) *Purification:* in step, check the purity by using ion-exchange chromatography and remove the impurity.

(iv) *Inactivation:* optimum concentration of formaldehyde is used for inactivation of viruses.

Quality by design approach in the coating process quality cannot be tested into the product but it should be built into the product. The quality by design approach can explain the coating process (Figure 4.3) important aspect that affects the coating process is given below.

Formulation Consideration	Process Parameter	Equipment Design

• Coating liquid
  • Type of solvent
  • Type of coating polymer
  • Concentration of coating polymer
  • Concentration of plastisizer
  • Viscosity of the polymer solution
  • Others
• Core material
  • Shape, size, porosity Density
  • Wetting property
  • Moisture content
  • Compatibility with polymer solution
• API solubility and compatibility with excipient
	• Inlet air temperature
	• Atomisation pressure
	• Air flow rate for fludization
	• Product bet temperature
	• Spray rate of solution polymer
	• Humidity of inlet and outlet air
		• Type and size
		• Nozzel
		• type/numbers/design/location
		• Air distribution plate type and pore size
		• Pump type

Fig. 4.3 QbD Approach for Coating Process

Six Sigma

Six Sigma was developed at Motorola in the 1980s as a method to measure and improve high-volume production processes. Its overall goal was to measure and eliminate waste by attempting to achieve near-perfect results. The term six sigma refers to a statistical measure with no more than 3.4 defects per million. Numerous companies, including General Electric, Ford, and DaimlerChrysler, have credited six sigma with saving them billions of dollars.

Six sigma is a statistically oriented approach to process improvement that uses various tools, including statistical process control (SPC), total quality management (TQM), and design of experiments (DOE). It can be coordinated with other major initiatives and systems, such as new product development, materials requirement planning (MRP), and just-in-time (JIT) inventory control. The Six Sigma concepts is to identify critical quality attributes in a process and then control the process by reducing variations, improving process capabilities, increasing stability, and designing systems to sustain the improvements. Six sigma was initially thought of as a system that could be used only in manufacturing operations, but more recently it is successful in nonmanufacturing processes as well, such as accounts payable, billing, marketing, and information systems.

At first glance, six sigma might seem too structured to be effective in analyzing processes not standard and repetitive as in manufacturing situations, but the theory of six sigma is flexible enough to suit any process. Nevertheless, many lessons learned on production lines are very relevant to other processes as well. The following is a brief description of the steps involved in the six sigma process:

1. Break down business process flow into individual steps
2. Define what defects there are
3. Measure the number of defects
4. Probe for the root cause
5. Implement changes to improve
6. Re-Measure
7. Take a long-term view of goals

Variations to a process can be calculated using the Process Capability Index (Cp). Cp is a measure of the ability of the process to produce outputs that meet the specification.

$$Cp = (USL - LSL)/6\sigma$$

Where

USL = Upper specification limit

LSL = Lower specification limit

Sigma = Standard deviation

When

Cp< 1 means the process is unacceptable

Cp= 1 means the process is acceptable

Cp> 1 means the process is desirable

The Six Sigma method to control process variability is the Define-Measure-Analyze-Improve-Control,

(DMAIC) model:

- **Define** – to determine the objectives and scope of the project to control process variability

- **Measure** – to identify one or more products, map the process, assess process performance, estimate baseline capability, and quantify problems

- **Analyze** – to evaluate and reduce the variables and determine the root causes

- **Improve** – to discover variable relationships, establish operating tolerances, and validate measurements

- **Control** – to determine the ability to control the vital few factors and implement process control systems.

The outcome is a control plan, which includes

- *Training plan* – training process owner and operators; use of control charts; understanding and using implemented documentation; knowing the response plan and able to use it if necessary.

- *Documentation plan* – new process step/s are integrated into normal processes and systems, procedures, policies, are modified to sustain improvement.

- *Monitoring plan – the team must document and monitor the process using metrics defined, evaluate the solution, assess the capability of the process over time and establish control systems for a long-term solution.*

- *Response plan* – establish checkpoints to signal out-of-control conditions and actions to undertake.

- *Institutionalization plan* – align systems and structures, develop standards and procedures, and communicate with all stakeholders.

Six Sigma implementation starts with the formation of a team to investigate and control the variability in processes.

The team would consist of the following:

- *Champions* – responsible for promoting the Six Sigma methodology throughout the company

- *Master Black Belts* – responsible for teaching executive managers and champions the basics of Six Sigma and helping to select the right people for the Six Sigma Project; also mentors Black and Green Belts

- *Black Belts* – responsible for leading Six Sigma project teams full time

- *Green Belts* – responsible for helping the Black Belts in their functional area, working part-time on a project

Standardized Systems/ISO Standards

The International Organization for Standardization (ISO), its central office located in Geneva, is the world's largest developer and publisher of international standards. It forms a platform for several countries to establishing quality systems. It has a network of National Standards Institutes in 163 participating countries, including the Bureau of Indian Standards (BIS) of India. The ISO international standards are highly specific to product, material, and process, ensuring the regulation from start to finish.

History and origins of ISO

The International Standardization Organization (ISO) was formed in 1947 as a non-governmental organization to promote, the development of standards to facilitate the international exchange of goods and services. It will also seek international co-operation in scientific, technological, and economic activities. ISO consists of about 100 countries as members although its number keeps growing and these member nations are mostly represented in ISO by their national standards organizations. ISO issued as a short form for the International Organization for Standardization. The term ISO is derived from the Greek word 'isos', which means 'equal,' thus the use of equal standards to guide the international exchange of goods and services.

As defined by the Central Secretariat of ISO, the aim of international standardization of facilitating trade exchange and technology transfer can be achieved through the following:

- Enhanced product quality and reliability at a reasonable price;
- Improved health, safety and environmental protection and reduction of waste;
- Greater compatibility and interoperability of goods and services;
- Simplification for improved usability;
- Reduction in the number of models and thus, reduction in costs;
- Increased distribution efficiency and ease of maintenance.

International standards serve a critical function in today's global markets. They remove significantly, as ISO notes, 'technical trade barriers.' Such barriers could be expected when there is no uniformity in standards among similar industries and technologies around the world. Increasingly, nations in the same regions are aligning themselves to protect their market territory and without standardization of industrial practices, it will be extremely difficult to compete outside one's regions. International standardization is expected to continue growing in its popularity worldwide. We adapt the reasons provided by the Central Secretariat of ISO as follows.

ISO 9000 Family Standards

ISO 9000 is a series of quality management systems (QMS) standards created by the International Organization for Standardization, a federation of 132 national standards bodies. The ISO 9000 QMS standards are not specific to products or services but apply to the processes that create them. An organization that would like to have ISO certification should meet all the criteria stated in the ISO standards and pass a detailed audit conducted by an ISO auditor. In some industries, ISO certification has become necessary; for example, some large manufacturers require all suppliers to be ISO certified. While ISO certification is highly respected, if it is not a trend in your specific industry, the additional cost of certification is a deterrent to most managers. It is very possible to reach the desired quality level within an organization with a well-planned quality system and without going through all additional steps for ISO certification. QS-9000, released in 1994, is the ISO 9000 derivative for suppliers to the automotive Big Three: DaimlerChrysler, Ford, and General Motors. This quality management system standard contains all ISO 9001:1994, along with automotive sector-specific, Big Three, and another original equipment manufacturer (OEM) customer-specific requirements.

- **ISO 9000** is a set of international standards on quality management and quality assurance developed to help companies effectively document the quality system elements to be implemented to maintain an efficient quality system. They are not specific to any one industry and can be applied to organizations of any size. ISO 9000 can help a company satisfy its customers, meet regulatory requirements, and achieve continual improvement. However, it should be considered as a first step, the base level of a quality system, not a complete guarantee of quality. These standards have been revised periodically, in 1994, 2000, and 2008, 2015.

- **ISO 9000:2015: Quality management systems - Fundamentals and vocabulary** (definitions)

- **ISO 9001:2015: Quality management systems - Requirements**

- **ISO 9004:2009:** Quality management systems – Manage of the sustained success of an organization (continuous improvement)

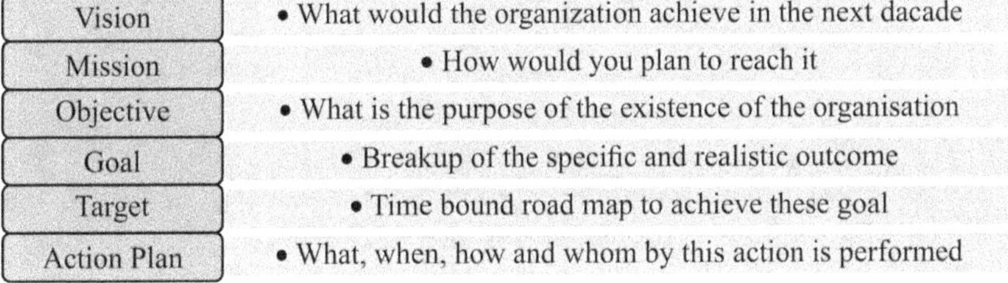

Fig. 4.4 Vision based activities per ISO 9001

ISO 9000 series:

ISO 9001:1987—Model for quality assurance in design, development, production, installation, and servicing. This is for companies and organizations whose activities include the creation of new products.

ISO 9002:1987—Model for quality assurance in production, installation, and servicing has the same material as ISO 9001, but without covering the creation of new products.

ISO 9003:1987—Model for quality assurance in final inspection and test.

ISO 9004:2009—Introduced in 2009, this focuses on how to make a quality management system more efficient and effective.

ISO 19011:2011—Introduced in 2011, this sets out guidance on internal and external audits of quality management systems.

ISO 10006—Quality management—guidelines for quality management in projects

ISO 10007—Quality management—guidelines for configuration management

ISO 10012—Quality assurance requirements for measuring equipment

ISO 10013—Quality management—guidelines for developing quality manuals

ISO 10014—Quality management—guidelines for economic effects of quality

ISO 10015—Quality management—guidelines for continuing education and training

ISO 9000 certification

Individuals and organizations **cannot** be certified to ISO 9000. ISO 9001 is the only standard within the ISO 9000 family to which organizations can certify.

ISO 9000:2000

ISO 9000:2000 refers to the ISO 9000 update released in 2000. The Technical Committee responsible for the ISO 9000 family developed specifications for the ISO 9000:2000 revisions, leading to a significant advancement of the standards and reflecting contemporary concepts of quality management.

The ISO 9000:2000 revision has five goals:

Meet stakeholder need

Be usable by all sizes organizations

Be usable by all sectors

Be simple and clearly understood

Connect quality management systems to business processes

(From ISO 9000:2000 Shifts Focus of Quality Management System Standards, by Jack West.)

ISO 9000:2000 was again updated in 2008 and 2015. ISO 9000:2015 is the most current version.

ISO 9000 principles of quality management

The ISO 9000:2015 and ISO 9001:2015 standards are based on seven quality management principles that senior management can apply for organizational improvement:

1. *Customer focus*

   - *Understand the needs of existing and future customers*
   - *Align organizational objectives with customer needs and expectations*
   - Meets customer requirements
   - Measure customer satisfaction
   - Manage customer relationships
   - Aim to exceed customer expectations
   - Learn more about the customer experience and customer satisfaction.

2. *Leadership*

   - Establish a vision and direction for the organization
   - Set challenging goals
   - Model organizational values
   - Establish trust
   - Equip and empower employees
   - Recognize employee contributions

3. *Engagement of people*

   - Ensure that people's abilities are used and valued
   - Make people accountable
   - Enables participation in continual improvement
   - Evaluate individual performance
   - Enable learning and knowledge sharing
   - Enable open discussions of problems and constraints

4. *Process approach*

   - Manage activities as processes
   - Measure the capability of activities

- Identify linkages between activities
- Prioritize improvement opportunities
- Deploy resources effectively

5. *Improvement*

- Improve organizational performance and capabilities
- Align improvement activities
- Empowers people to make improvements
- Measure improvement consistently
- Celebrate improvements
- Evidence-based decision making
- Ensure the accessibility of accurate and reliable data
- Appropriate methods are used to analyze data
- Make decisions based on analysis
- Balance data analysis with practical experience

6. *Relationship management*

- We identify and select suppliers to manage costs, optimize resources, and create value
- Establish a relationship considering both the short and long term
- Share expertise, resources, information, and plans with partners
- Collaborate on improvement and development activities
- Recognize supplier successes

ISO 14000 Family Standards

During the 1992 Earth Summit in Rio de Janeiro, the Business Council for Sustainable Development suggested that the International Organization for Standardization (ISO), which had already established standards for the quality of air, water, and soil, should develop international standards for environmental performance based on the concept of sustainable development. In 1993, ISO formed the technical Committee 207 on Environmental Management to develop international standards for environmental management tools and systems.

A set of international standards brings a worldwide focus to the environment, thus encouraging a cleaner, safer, healthier world for us all. The existence of the standards allows organizations to focus on environmental efforts against internationally accepted criteria. The ISO

14000 series of standards developed by ISO/TC 207 effectively addresses the needs of organizations worldwide by providing a common framework for managing environmental issues. They promise to effect a broadly based improvement in environmental management, which in turn can facilitate trade and improve environmental performance worldwide.

The primary objective of the ISO 14000 series of standards is to promote effective environmental management systems in organizations. The standards seek to provide cost-effective tools that use best practices for organizing and applying information about environmental management. The ISO 14000 series embodies a new approach to environmental protection for organizations in the global marketplace.

It challenges an organization to;

- Take stock of its impact on the environment

- Establish its objectives and targets

- Commit itself to effective and reliable processes, prevention of pollution, and continual improvement

- Bring all employees and managers into a system of shared and enlightened awareness and personal responsibility for the organization's performance concerning the environment.

The ISO 14000 standards follow the Plan-Do-Check-Act (PDCA) strategy, a self-improving loop that includes the following steps:

- *Plan:* Launching of a confirmed policy by the management planning of objectives concerning this policy

- *Do:* Implementation of the provisions specified in the plan

- *Check:* Verification and assessment of results and progress achieved

- *Act:* Review the continual improvement of the system.

Although the ISO 14000 standards are designed to be mutually supportive, they can also be used independently of each other to achieve environmental goals.

ISO 14000 Series

ISO 14001—environmental management system. Specifications with guidance for use. This serves as the basis for certification

ISO 14004—environmental management system—general guidelines on principles, systems, and supporting techniques.

ISO 14006—environmental management systems—guidelines for incorporating ecodesign

ISO 14010—guidelines for environmental auditing, general principles.

ISO14011—guidelines for environmental auditing, audit procedures, auditing of environmental systems

ISO14012—guidelines for environmental auditing—qualification criteria for environmental auditors.

ISO 14014—initial review.

ISO 14015—site assessment guidelines.

ISO 14020, ISO 14021, ISO 14022, ISO 14023, and ISO 14024—basicallyfor environmental labeling.

ISO 14030—discusses post-production environmental assessment.

ISO 14031—environmental performance evaluation.

ISO 14040, ISO 14041, ISO 14042, ISO 14043—for life cycle assessment.

ISO 14046—sets guidelines and requirements for water footprint assessments of products, processes, and organizations. Includes only air and soil emissions that impact water quality in the assessment.

ISO 14050—terms and definitions.

ISO 14060—product standards.

ISO 14062—discusses making improvements to environmental impact goals.

ISO 14063—environmental communication—Guidelines and examples.

ISO 14064—measuring, quantifying, and reducing greenhouse gas emissions.

ISO19011—single audit protocol for both 14000 and 9000 series standards together.

The benefit of ISO 14000 standards

The whole ISO 14000 family provides management tools for organizations to control their environmental aspects and to improve their environmental performance. Together, these tools can provide significant economic benefits, including

- Reduced raw materials/resource use;

- Reduced energy consumption;
- Improved process efficiency;
- Reduced waste generation and disposal costs;
- Utilization of recoverable resources;
- Potentially lower insurance premiums.

The ISO 14000 series of standards allows small- and medium-sized enterprises (SMEs) to compete on a level playing field with larger and even global companies, as the same standard applies to all types and sizes of organizations.

Certification

After an organization has implemented an EMS tailored to ISO 14001 and operates it for an adequate period (generally three to six months, and on satisfactory completion of internal audit) a third-party certification body

- Filling an application can be contracted to verify that the organization's EMS meets the standard's requirements and that the environmental objectives and goals are being implemented at all levels of the operation. This is the external audit process. Third-party certification includes the following basic steps:
- Manual or policy review
- Pre-assessment (optional)
- Certification
- Surveillance

A thorough review of an organization's environmental management system documentation is required under ISO 14001. In most organizations, documentation is structured in the following three-tier hierarchy:

- The EMS manual and/or environmental policy;
- Operating procedures; and
- Environmental records.

During an EMS audit, documents are reviewed to determine whether they meet the applicable requirements of the ISO 14001 standard or any other requirements. The criteria against which the EMS is being measured need to be clearly defined in the company's documentation. If this review shows that the organization's documented system is incapable of meeting the requirements, no more time should be spent on the audit until the situation is corrected.

During the audits, we come across non-conformances. If the non-conformance (NC) is identified as a "minor," it is a problem that can be easily corrected. Usually, a minor non-conformance is not something that will block the auditing process. When a "major" non-conformance is identified, it usually means that a significant change to the EMS has to be done, like adding a procedure or changing practice. Corrective action must be taken to eliminate the cause of the non-conformance. Organizations should seek to find the root cause of the non-conformance and prevent it from recurring. Once a major non-conformance is corrected, a follow-up assessment limited to the area of concern is usually required. As long as there are non-conformances, both minor and major, certification is impossible.

Once a company is certified, it receives a certificate and is listed in a register or directory published by the certification body. Most certifiers conduct surveillance visits once a year. The standards validity of a registration certificate is three years, after which a full re-audit is done for renewal. After certification, companies must continue to show that they are meeting the requirements of the standard. Should a company fall short of this goal, its registration may be revoked.

ISO 14001 certification requires compliance in four organizational areas:

- Implementation of an environmental management system;
- Assurance that procedures are in place to maintain compliance with laws and regulations;
- Commitment to continual improvement; and
- Commitment to waste minimization and prevention of pollution

Integration of ISO 14000 with ISO 9000

Both ISO 14001 and ISO 9001 deal with management systems that form an integral part of the overall management of an organization. Both systems are designed to contribute to improving the business performance of an organization. The standards share many common system elements which can be easily used to integrate the environmental management system with the quality management system of an organization in the following areas

- **Documentation**

  Both standards require the development of documentation to address their requirements. One possible method for integrating documents is to develop separate policy manuals but to prepare common documented procedures for various QMS and EMS requirements.

- *Management responsibility*

  Both standards require the commitment of top management to establish and implement management systems.

- *Resource management*

  Both systems contain requirements for resources, including human resources, for applying the policies and achieving the objectives of the organization. Requirements for competence, awareness, and training are common to both standards.

- *Product realization*

  The ISO 9001 quality management system addresses various product realization processes separately, i.e. Planning of product realization, customer-related processes, design and development, purchasing and production, and service provision. In these processes, the identification and/or operational control of significant environmental aspects may be integrated.

- *Measurement, analysis, and improvement*

  Most of the requirements in this section (internal audits, monitoring, and measurement, control of non-conformities, corrective action, preventive action, continual improvement) are common to both the EMS and QMS. These common requirements can be addressed simultaneously in the management system procedures.

Certification bodies are also agreeing to undertake joint certification audits in companies that have integrated an EMS and a QMS, thus reducing the third-party certification. ISO 19011 "Guidelines on quality and/or environmental auditing systems" provides guidelines for quality and/or environmental management system auditing. The benefits of ISO 19011 can be summarized as follows

- Better applicability to the conduct of internal audits, and more focused on use by small- and medium-sized enterprises

- More flexible approach to auditor qualifications and audit team selection; and

- Applicability to combined audits, and herewith bridging the gap between quality and environmental management tools.

National Accreditation Board for Testing Laboratory (NABL)

The concept of laboratory accreditation was developed to provide third-party certification that a laboratory is competent to perform a specific test or type of test.Laboratory accreditation is a means to improvecustomer confidence in the test reports issued by thelaboratory so that the clinicians and through them thepatients shall accept the reports with confidence. The National Accreditation Board for Testing and calibration Laboratories (NABL) is an autonomousbody under the aegis of the Dept. of Science &Technology, Govt. of India, and is registered under the Societies Act. NABL, which was initially establishedto provide accreditation to testing &calibration laboratories, later on, extended its servicesto the clinical laboratories in our country.

Govt. of India has authorized NABL as the sole accreditation body for testing and calibrationlaboratories. The objective of NABL is to provide a third-party assessment of quality and technical competence. Four years ago NABL established links withinternational bodies - Asia Pacific LaboratoryAccreditation Cooperation and the International Laboratory Accreditation Cooperation. This has impacted international recognition of NABL accreditedlaboratories. The international standard currentlyfollowed by NABL is ISO 15189, specific for medicallaboratories. The National Accreditation Board for Testing and Calibration Laboratories (NABL) is an accreditation body, with its accreditation system established following ISO/IEC 17011. "Conformity Assessment –Requirements for Accreditation bodies accrediting conformity assessment bodies." NABL provide voluntary accreditation services to:

- Testing Laboratories Following the ISO/IEC 17025 'General Requirements for the Competence of Testing and Calibration Laboratories

- Calibration laboratories follow the ISO/IEC 17025 'General Requirements for the Competence of Testing and Calibration Laboratories

- Medical testing laboratories following ISO 15189 'Medical laboratories -Requirements for quality and competence'

- Proficiency Testing Providers (PTP) following ISO/IEC 17043 "Conformity assessment — General requirements for proficiency testing" and

- Reference material producers (RMP) following ISO 17034 "General requirements for the competence of reference material producers."

NABL is Mutual Recognition Arrangements (MRA) signatory to ILAC as well as APAC for the accreditation of Testing and Calibration Laboratories (ISO/IEC 17025), Medical Testing Laboratories (ISO 15189), Proficiency Testing Providers (PTP) (ISO/IEC 17043) and Reference material producers (RMP).

NABL Vision-To be the world's leading accreditation body and to enhance stakeholders' confidence in its services.

NABL Mission-To strengthen the accreditation system accepted across the globe by providing high quality, value-driven services, fostering APAC/ILAC MRA, impaneling competent assessors, creating awareness among the stakeholders, initiating new programs supporting accreditation activities, and pursuing organizational excellence.

Organization Structure -NABL is one of the constituent boards of the Quality Council of India (QCI), an autonomous body under the Department for Promotion of Industry & Internal Trade (DPIIT), Ministry of Commerce and Industry, Government of India. Organization structure is shown below in figure 4.5.

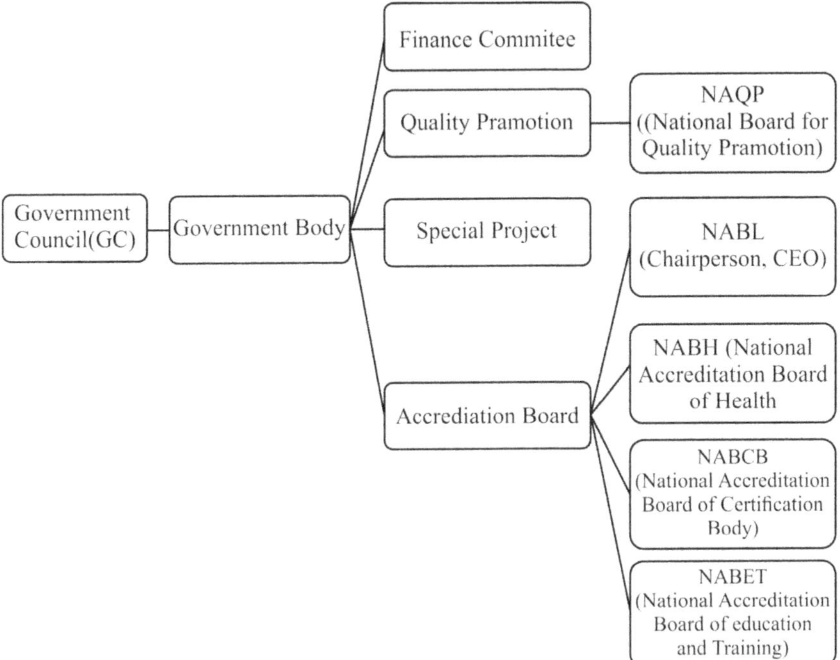

Fig. 4.5 Organisation structure

Composition of the NABL Board

Table 4.3 Composition of NABL boards

1	Chairperson
2	Representative of National Standards Body (Ex-Officio) – DG, BIS
3	Representative of National Metrological Institute (Ex-Officio)– Director, NPL
4	A representative from Govt. / Regulators [5] (Ex-Officio – Adl. Secretary-DPIIT, Secretary-DST, Director-General-DGHS), Secretary-cum-Scientific Director-Indian Pharmacopoeia Commission (IPC), Additional Secretary
5	A representative from Associations [5] (Ex-Officio – DG-CII, SG-FICCI, SG-ASSOCHAM), Automotive Research Association of India, Electrical Research and Development Association (ERDA)
6	A representative from Industries [5] – Vice President (R&D)-MRF Ltd, Director (IS&P)-BHEL, Vice President (Technology & New Material Division)-TATA Steel, Executive Director (QC) -Indian Oil Corporation Limited (IOCL)
7	A representative from Laboratories [4] – Electronics Regional Test Laboratory (ERTL), Secretary-Textile Committee, Director General-Central Power Research Institute (CPRI), Fare Labs Pvt Ltd.
8	NABL Accreditation Committee Chairman
9	Quality Council of India – Secretary General (Ex-Officio)
10	Member Secretary – CEO, NABL (Ex-Officio)

Scope of Accreditation- NABL Accreditation is currently given in the following fields and disciplines. The multi-disciplinary CABs shall have to apply the relevant discipline separately depending upon to which discipline the scope belongs.

Table 4.4 Scope of NABL accreditation

Testing laboratories	Calibration laboratories	Medical laboratories
<ul><li>Chemical</li><li>Biological</li><li>Mechanical</li><li>Electrical</li><li>Electronics</li><li>Fluid Flow</li><li>Forensic</li><li>Non-Destructive (NDT)</li><li>Photometry</li><li>Radiological</li><li>Diagnostic Radiology QA Testing</li><li>Software and IT System</li></ul>	<ul><li>Mechanical</li><li>Electro-Technical</li><li>Fluid Flow</li><li>Thermal</li><li>Optical</li><li>Medical Devices</li><li>Radiological</li></ul>	<ul><li>Clinical Biochemistry</li><li>Clinical Pathology</li><li>Hematology</li><li>Microbiology and infectious disease serology</li><li>Histopathology</li><li>Cytopathology</li><li>Flow Cytometry</li><li>Cytogenetics</li><li>Molecular Testing</li></ul>

Table 4.4 *Contd....*

MEDICAL IMAGING-CONFORMITY ASSESSMENT BODIES (MI-CAB)
- Projectional Radiography & Fluoroscopy
  - X-Ray, Bone Densitometry (DEXA), DentalX-Ray-OPG, Mammography, etc.
  - Fluoroscopy Computed Tomography (CT)
- Magnetic Resonance Imaging (MRI)
- Ultrasound and Colour Doppler
- Nuclear Medicine
  - SPECT
  - PET CT
  - PET MRI
- *Basic Diagnostic Interventional Radiology Procedures

(Image-guided Core Biopsy and/or Needle Aspiration e.g. Fine Needle Aspiration Cytology)
**For only such IR procedures that will be carried out by Radiologists*

PROFICIENCY TESTING PROVIDERS	**REFERENCE MATERIAL PRODUCERS**
- Testing - Calibration - Medical - Inspection	- Chemical Composition - Biological and Clinical Properties - Physical Properties - Engineering Properties - Miscellaneous Properties

Getting Ready for Accreditation

A laboratory should make a definite plan for obtaining accreditation and nominate a sponsible person as QUALITY MANAGER (who should be familiar with the laboratory's existing quality system) to coordinate all activities related to seeking accreditation. The laboratory should carry out the following important tasks toward getting ready for accreditation:

1. Contact the NABL Secretariat with a request for procuring relevant NABL documents.

2. Get fully acquainted with all relevant documents and understand the assessment Procedure and methodology of making an application.

3. Train a person on Quality Management System and Internal Audit (4-day residential training courses conducted by NABL.

4. Prepare QUALITY MANUAL as per the ISO 15189standards.

5. Standard operating procedures for each investigation carried out in the laboratory.

6. Ensure effective environmental conditions (temperature, humidity, storage placement, etc.).

7. Ensure calibration of instruments/equipment. Only NABL ACCREDITED CALIBRATIONLABORATORIES are authorized

to provide calibration. The NABL website gives the names of NABL accredited calibration laboratories in the various fields of Accreditation.

8. Impart training on the key elements of documentation, such as document format, authorization of document, issue, and withdrawal procedures, document review and change, etc. Each document should have ID No. Name of controlling authority, the period of retention, etc.

9. Ascertain the status of the existing quality system and technical competence of NABL standards and address the question "Is the system documented and effective OR does it need modification?"

10. Remember Quality Manual is a policy document, which has to be supplemented by a set of other next-level documents. Therefore ensure that these documents are well prepared.

11. Ensure proper implementation of all aspects that have been documented in the Quality Manual and in other documents.

12. Incorporate Internal Quality Control (IQC) practice while patients' samples are analyzed.

13. Document IQC data as well as the uncertainty of measurements. Maintain Levy Jennings charts.

14. Participate in External Quality Assessment Schemes (EQAS).If this is not available for certain analytes, participate in inter-laboratory comparison through the exchange of samples with NABL accredited laboratories.

15. Document corrective actions on IQC/EQA outliers.

16. Conduct Internal Audit and Management Review.

17. Apply to NABL along with the appropriate fee.

Accreditation Process

An applicant laboratory is expected to submit to NABL5 copies of the application and 5 copies of the Quality Manual.

- The Quality Manual will be forwarded by NABL to the Lead Assessor to judge the adequacy of the quality manual and whether it complies with ISO 15189 standards.

- Thereafter the Lead Assessor will conduct a Pre-Assessment of the laboratory for one day. Based on the Pre-Assessment report, the laboratory may have to take certain corrective actions, to be fully prepared for the final assessment.

- It is essential for the applicant and accredited laboratories to satisfactorily participate in Proficiency testing/Inter-laboratory comparisons/External quality assessment programmed as the Asia Pacific Laboratory Accreditation Cooperation (APLAC) Mutual Recognition Arrangement calls for mandatory participation in such programs.

- Finally, when the laboratory is ready, the Lead Assessor and a team of technical assessors will conduct the final assessment. The number of technical assessors will depend on the number of disciplines applied. The accreditation process involves a thorough assessment of all elements of the laboratory that contribute to the production of accurate and reliable test data. These elements include staffing, training, supervision, quality control, equipment, recording and reporting of test results, and the environment in which the laboratory operates. The laboratory may have to take certain corrective actions, after the final assessment.

- After satisfactory corrective actions are taken at the laboratory (within 3 months), the accreditation committee will examine the report and if satisfied recommend accreditation.

The time required for the process of accreditation will depend upon the preparedness of the laboratory and its response to the non - conformances raised during the pre-assessment and final assessment. The total duration ranges between 6 and 8 months.

International Recognition- To ensure the acceptance of test/calibration results issued by the accredited conformity assessment bodies (CAB) across the borders, NABL maintains linkages with the international bodies:

- International Laboratory Accreditation Co-operation (ILAC) and

- Asia Pacific Accreditation Co-operation (APAC).

NABL is a full member of ILAC and APAC and regularly takes part in their meetings. NABL is a Mutual Recognition Arrangement (MRA) signatory to ILAC as well as APAC for the accreditation programs – Testing and Calibration (ISO/IEC 17025), Medical (ISO 15189), Proficiency Testing Providers (PTP) (ISO/IEC 17043), and Reference material producers (RMP) (ISO 17034).

Surveillance and Re-Assessment

Accreditation into a laboratory shall valid for three years. NABL shall conduct annual surveillance of the accredited laboratories. The laboratories may enhance or reduce the scope of accreditation during surveillance. The laboratories should apply for renewal of accreditation, at least six months before the expiry of the validity of accreditation for which are-assessment shall be conducted.

Good Laboratory Practices (GLP)

GLP originated in the mid-1970s in the United States, when various deficiencies were revealed in contract laboratories during an FDA inspection. These deficiencies included, among others, the conduct of studies, record keeping, archiving, and animal husbandry. As a response, the chemical industry suggested the implementation of a quality management system, which was later named GLP. The FDA adopted GLP regulations in 1979. The Organisation for Economic Cooperation and Development (OECD) issued GLP guidelines in 1981 for chemicals. A set of rules, operating procedures, and the proper practices ensure the reliability of the data generated by pharmaceutical control laboratories.

Principles of Good Laboratory Practices:
1. Organization and personnel
2. Installations and facilities
3. Documentation
4. Equipment and instruments
5. Materials and reagents
6. Reference and assay samples
7. Assay methods. Validation
8. Self-inspections and audits
9. Quality assurance for the assays

Basic requirements of Good Laboratory Practices:
1. Norms or procedures as basic documents to establish quality standards and evaluate whether the manufactured product conforms to the specifications.
2. Sampling, inspections, and testing of materials are performed by trained personnel, using appropriate methods and following established norms and procedures.
3. The samples are taken by Quality Control personnel, following approved procedures.
4. The assay methods were validated.

5. The registries and record books are kept in such a way that no data can be entered unless the described sampling or inspection/assay procedures have been performed. Any deviation will be carefully recorded and investigated.

6. The finished products comply with the stated specifications; and are correctly bottled and labelled.

7. The evaluation of the product includes a revision and assessment of the documentation of the process and the evaluation of any deviations.

8. The certification by authorized personnel, following the previously specified procedures and requirements, is required before releasing a batch of the product.

9. Retained samples are stored following established procedures, ensuring that they are properly labelled and are kept under the specified storage conditions.

10. The quality control unit is separated from the production unit.

11. The Quality Control personnel will be granted access to the manufacturing facilities to carry out their job.

12. The Quality Control Unit must remain under the authority of a qualified and competent person and will have one or more laboratories equipped with all resources needed to ensure the reliability of the decisions made by Quality Control.

A. Multiple Choice Questions

1. Quality Management systems include
   a) Data management
   b) Customer satisfaction with product quality
   c) Quality analysis
   d) All

2. The definition of Quality Risk Management (QRM) has been mentioned in the ICH guideline
   a) Q8
   b) Q9
   c) Q10
   d) Q12

3. What is the full form of GMP?
   a) Good manufacturing practice
   b) General manufacturing practice

    c)   General manufacturing process

    d)   Good manufacturing process

4.   GMP facility is a major inspection carried out in?

    a)   1 year   b)   2 year

    c)   3 year   d)   5 year

5.   Current Good Manufacturing Practice for Finished Pharmaceuticals is included in the part of CFR?

    a)   CFR 210   b)   CFR211

    c)   CFR610   d)   CFR 600

6.   Stability testing of biotechnological/biological products includes which ICH guideline

    a)   Q5C   b)   Q1A

    c)   Q1B   d)   Q1C

7.   Good manufacturing practice guide for active pharmaceutical ingredients included in

    a)   Q7   b)   Q8

    c)   Q9   d)   Q10

8.   Electronic Common Technical Document (eCTD) is included in

    a)   M4   b)   M2

    c)   M8   d)   M3

9.   Validation of the process by taking the already available data of the manufacturing process is considered as

    a)   Prospective   b)   Concurrent

    c)   Retropespective   d)   Revalidation

10. The QBD approach helps to formulators for manufacturing formulation by

    a)   To minimize the variability   b)   To minimize the variability

    c)   Reduce the batch failure   d)   all

11. QBD term in full form is

    a)   Quality by Design   b)   Quality by drug

    c)   Quantity by design   d)   Qualification by design

12. QBD is defined in which ICH guidelines

    a)   Q8   b)   Q7

    c)   Q9   d)   Q10

13. Six Sigma process involves the following
    a) Define what defects there are
    b) Measure the number of defects
    c) Probe for the root cause
    d) All

14. ISO series for quality standards for manufacturing is included in
    a) ISO 9000   b) ISO 14000
    c) Both A and B   d) None

15. ISO series 14000 is considered for quality in
    a) Manufacturing
    b) Environment protection and performance
    c) Both
    d) none

16. NABL Medical testing laboratories following ISO for 'Medical laboratories -Requirements for quality and competence'
    a) ISO 15189   b) ISO 17025
    c) ISO1703   d) Both b &c

17. GLP in full form is
    a) Good laboratory practice
    b) General laboratory practice
    c) Good laboratory performance
    d) General laboratory performance

18. Good laboratory practices include
    a) Conduct of studies   b) Recordkeeping,
    c) Animal husbandry   d) All

19. Safety Pharmacology Studies for human pharmaceuticals are carried out according to which ICH guideline?
    a) S7A   b) S8
    c) S6   d) All

20. NABL laboratory certificate help to the organization
    a) International recognition   b) Rise in business
    c) Set benchmarks for testing laboratory
    d) ALL

B. Short Questions

1. Write a short note Total Quality Management

2. What is NABL? Explain different laboratory and their function.

3. Write a short note on GLP.

4. Write down the types of validation and the validation process for the analytical method of validation.

5. What is the difference between the QBD approach and the traditional approach in the pharmaceutical product development process?

C. Long Questions

1. Write a brief note on the ICH guidelines. Explain the quality guidelines in detail.

2. Explain the principles of the TQM system.

3. Write a note about the six sigma concepts.

5. Write the basic principles of ISO 9000. Explain the ISO 9000 series in detail.

6. What do you mean by QBD? Give an example and explain it in detail.

MCQs Answers

1	2	3	4	5	6	7	8	9	10	11	12	13	14	15	16	17	18	19	20
D	B	A	B	B	A	A	C	C	D	A	A	D	A	B	A	A	D	A	D

CHAPTER 5

Indian Regulatory Requirements

> Central Drug Standard Control Organization (CDSCO)

> State Licensing Authority: Organization

> Responsibilities

> Common Technical Document (CTD)

> Certificate of Pharmaceutical Product (COPP)

> Regulatory Requirements

> Approval Procedures for New Drugs

Introduction

Until the 20th century, the drug manufacturing industry in India was at its primary stages. The majority of drugs were imported from other countries. After the First World War, the demand for drugs increased enormously, producing the manufacturing of spurious and substandard drugs on the market. In 1947—the year of India's independence—the pharmaceutical market was dominated by western MNCs. The supply of drugs into the market during this time depended upon 80%_90% import. Later, in 1960, the government initiated policies emphasizing the production of drugs domestically. During that time, many unscrupulous foreign manufacturers flooded the Indian market with adulterated and spurious drugs. Further, in response to the famous "Gigantic Quinine Fraud," the Government formed a drug inquiry committee under Sir Ram Nath Chopra, also known as the "Chopra Committee," whose recommendations were later presented as "The Drug Bill" and amended to the Drugs and Cosmetic Act of 1940 (D and C Act) and Drugs and Cosmetic rules of 1945. By 1990, India was self-sufficient in the production of formulations and nearly self-sufficient in the production of bulk drugs. Today, India is the world's fifth largest producer of bulk drugs, ranks 3rd in the word in terms of production volume, and 13th in domestic consumption value, serving almost 95% of the country's pharmaceutical needs.

Medicines in India are regulated by **CDSCO** - Central Drugs Standard Control Organization under the Ministry of Health and Family Welfare. Headed by the Directorate General of Health Services CDSCO regulates the Pharmaceutical Products through **DCGI** - Drugs Controller General of India at Chair. **The Drugs Control General of India (DCGI)** main authority for clinical trials, ensures standards, registers all imported drugs, new drugs, biologics and medical devices.

The regulatory authorities of India are enlisted below:

- Ministry of Health and Family Welfare (CDSCO)

- Ministry of Science and technology (ICMR/CSIR)

- Central Drug Standard Control Organization (CDSCO)

- Ministry of chemical and fertilizers (NPPA)

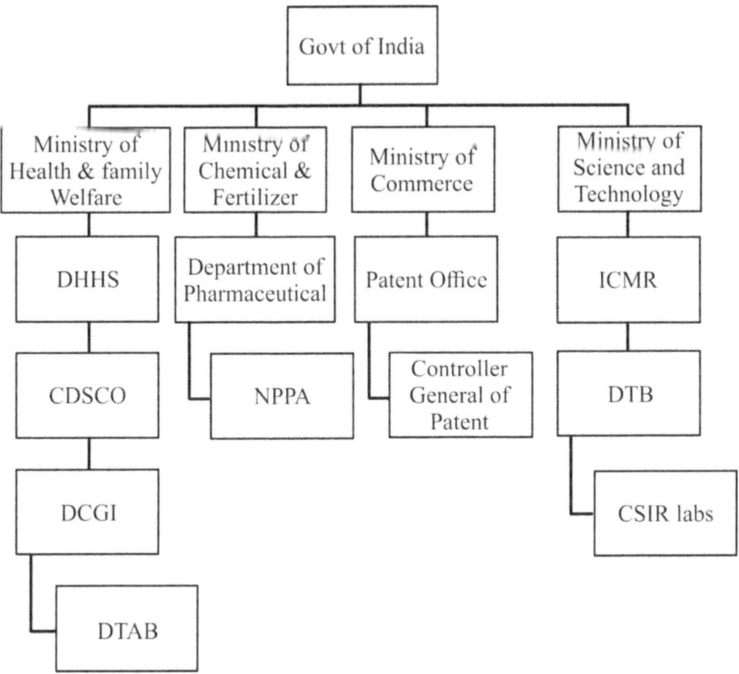

Fig. 5.1 Major Regulatory authority in India

The regulatory bodies involved in the pharmaceutical industry and their functions are as follows:

- **Indian Council of Medical Research (ICMR)**

The Indian Council of Medical Research (ICMR), is amongst the oldest medical research bodies in the world and the apex body in India for the formulation, coordination, and promotion of biomedical research, and is funded by the government of India through the Ministry of Health and Family Welfare. The Indian Research Fund Association was built in 1911, and was responsible of sponsoring and coordinating medical research in the country. After independence, some important changes were made in the organization and activities of the IRFA. It wasre-named as ICMR in 1949 with a remarkably broadened scope of functions.

- **Genetic Engineering Approval Committee (GEAC)**– Deals with genetic engineering and molecular biology, trials in which biotech products are used, will be referred to here by DCGI.

- **Department Of Biotechnology (DBT)**- Oversees the development of modern biology and biotechnology in India.

- **Atomic Energy Review Board (AERB)**- Has regulatory control over radiation equipment.

- **Baba Atomic Research Centre (BARC)**- Approves all radiation-related projects and Radio pharmaceuticals in India.

- **Drugs Consultative Committee (DCC)**- Provides technical guidance to the CDSCO.

- **Central Drugs Laboratory (CDL)**- Maintains quality control of drugs

- **Central License Approving Authority (CLAA)**- Provides approval for manufacturing licenses'

- **Drugs Technical Advisory Board (DTAB)**- Provides technical guidance to the CDSCO.

- **National Pharmaceutical Pricing Authority (NPPA)**- NPPA fixes or revises the prices of decontrolled bulk drugs and formulations and periodically updates the list under price control according to guidelines.

Under the current Indian legal and regulatory regime, the manufacture, sale, import, export and clinical research of drugs and cosmetics is governed by the following laws-

- The Drugs and Cosmetics Act, 1940

- The Pharmacy Act, 1948

- The Drugs and Magic Remedies (Objectionable Advertisement) Act, 1954

- The Narcotic Drugs and Psychotropic Substances Act, 1985

- The Medicinal and Toilet Preparations (Excise Duties) Act, 1956

- The Drugs (Prices Control) Order 1995 (under the Essential Commodities Act.

Some other laws have a bearing on pharmaceutical manufacture, distribution and sale in India. The important ones being

- The Industries (Development and Regulation) Act, 1951

- The Trade and Merchandise Marks Act, 1958

- The Indian Patent and Design Act, 1970

- Factories Act, 1948

Central Drug Standard Control Organization (CDSCO)

The Central Drug Standard Control Organization (CDSCO) is the national regulatory body that controls the manufacture, sale, import, export, repacking, etc. of pharmaceuticals and medical devices, under the Ministry of Health and Family Welfare, Government of India. It does similar functions as has been done by the European Medicines Agency of the European Union, the PMDA of Japan, the Food and Drug Administration of the United States, and the Medicines and Healthcare products Regulatory Agency of the United Kingdom. Every country across the world has a similar body.

There are other bodies such as the Drug Technical Advisory Board (DTAB) and the Drug Consultative Committee (DCC), which advise the DCGI. The CDSCO is divided into zonal offices, which primarily take care of the pre-licensing and post-licensing inspections, post-market surveillance, and recalls, if necessary. As per the Drugs and Cosmetics Act, 1940 and rules 1948 there under, the Central Government has some functions too. The Central Drugs Standard Control Organization (CDSCO) is the Central Drug Authority to carry out those functions. There are six zonal offices, four sub-zonal offices, 13 post offices, and 7 laboratories under the control of CDSCO. The headquarters of the Central Drugs Standard Control Organization is located at FDA Bhawan, Kotla Road, New Delhi 110002, and delivers its functions under the Directorate General of Health Services.

Responsibilities of CDSCO

Under the Drug and Cosmetics Act, the primary responsibilities of the Central Drug Control Authorities is to regulate the

1. Approval of the New Drugs,
2. Grant permission to conduct Clinical Trials in the country,
3. Formulation of the standards for Drugs,
4. Control over the quality of imported Drugs,
5. Coordination with the activities of State Drug Control Organisations, and
6. We provide advice to ensure uniform enforcement of the Drugs and Cosmetics Act throughout the country.

Organization

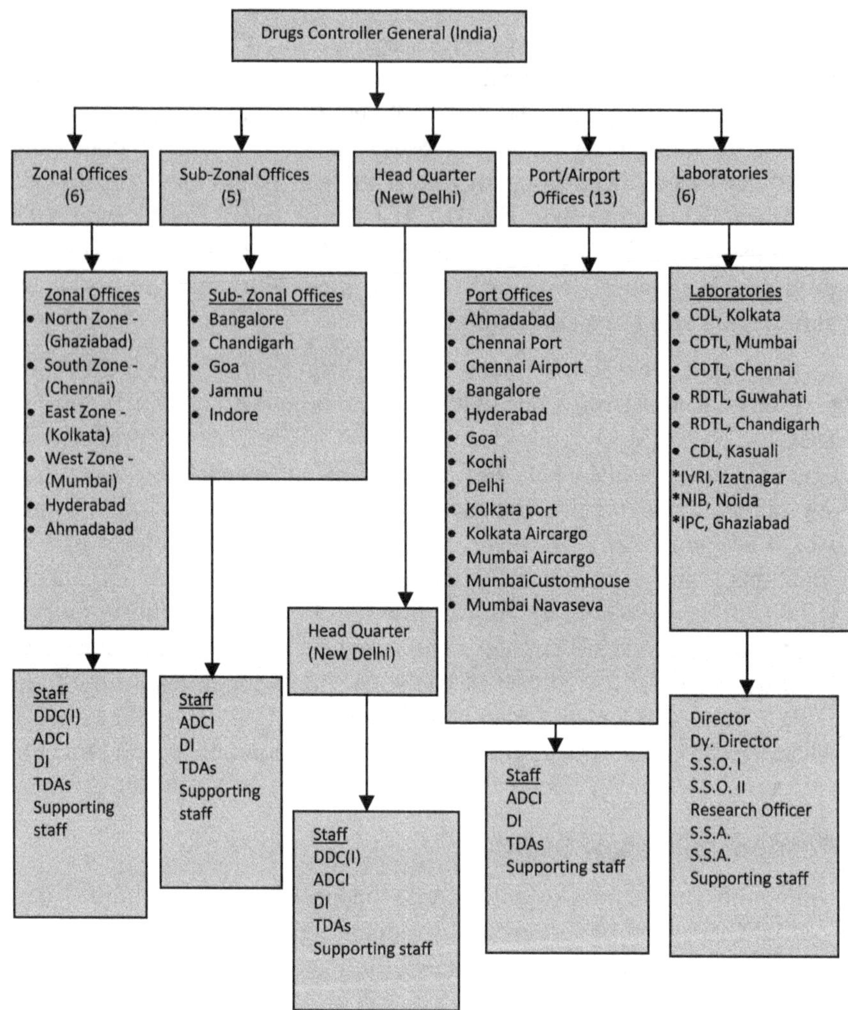

Fig. 5.2 Organisation of CDSCO

The Drug Controller General of India is responsible primarily for granting approval or issuing licenses to process or manufacture-specified categories of drugs such as blood and blood products, I.V. Fluids, Vaccines, and Sera.

Zonal Offices of CDSCO

Four zonal offices of the Central Drug Standard Control Organisation at Mumbai, Kolkata, Chennai, and Ghaziabad have been established by the Central Government. In close collaboration with the State Drug Control Administration, the zonal Offices work and assist them in securing uniform enforcement of the Drug Act and other related legislations, across the country (India).

Laboratory in India

Central Drugs Laboratory (CDL)

The Central Drugs Laboratory, Kolkata was established under the Drug & Cosmetics Act, of 1940. It is the national statutory laboratory of India's Government for quality control of Drugs and Cosmetics. It is the oldest quality control laboratory of the Drug Control Authorities in India. It functions under the administrative control of the Director-General of Health Services in the Ministry of Health and Family Welfare.

The functions of the Laboratory include

I. Statutory Functions

(a) Analytical quality control of the majority of the imported drugs available in the Indian market.

(b) Analytical quality control of drugs and cosmetics manufactured within the country on behalf of the Central and State Drug Control Administrations.

(c) Acting as an Appellate authority in matters of disputes relating to the quality of Drugs.

II. Other Functions

(a) Collection, storage, and distribution of International Standard, International Reference Preparations of Drug and Pharmaceutical Substances.

(b) Preparation of National Reference Standards and maintenance of such Standards. Maintenance of microbial cultures is useful in drug analysis, the distribution of standards and cultures to State Quality Control Laboratories and drug manufacturing establishments.

(c) Training of Drug Analysts deputed by the State Drug Control Laboratories and other Institutions.

(d) Training of World Health Organisation fellows from abroad on modern methods of Drug Analysis.

(e) Advise the Central Drug Control Administration regarding the quality and toxicity of drug awaiting license.

(f) To work the analytical specifications for the preparation of Monographs for the Indian Pharmacopoeia and the Homoeopathic Pharmacopoeia of India.

(g) To undertake analytical research on standardization and methodology of drugs and cosmetics.

(h) Analysis of cosmetics received as survey samples from the Central Drug Standard Control Organisation.

(i) Quick analysis of life-saving drugs on an all-India basis received under National Survey of Quality of Essential Drug Programme from Zonal Offices of Central Drug Standard Control Organisation.

In addition to the above functions, the Central Drug Laboratory actively collaborates with the World Health Organisation in the preparation of International Standards and Specifications for International Pharmacopoeia. It also undertakes a collaborative study on behalf of the Indian Pharmacopoeia Committee. The senior Officers of the Laboratory have been appointed as Government Analysts on behalf of most of the States of the Union for the analysis of drug samples.

Central Drugs Testing Laboratory (CDTL), Chennai

Central Drug Testing Laboratory is one of the four National Laboratories in India engaged in the research and analysis of Drug and Cosmetics as per the Drug and Cosmetics Act, 1940.

Central Drugs Testing Laboratory (CDTL), Hyderabad

Central Drug Testing Laboratory is one of the newly established laboratories in the state of Andhra Pradesh. It has been engaged in testing, research, and analysis of Drug and Cosmetics as per the Drug and Cosmetics Act, 1940.

Central Drugs Testing Laboratory (CDTL), Mumbai

The Central Drugs Testing Laboratory in Mumbai is another national statutory laboratory of India's Government. It works under the administrative control of the Drug Controller General (India), DGHS, Ministry of Health, and Family Welfare.

The major functions of the laboratory included

➢ Testing of imported bulk drugs and formulations referred by ADCs, Mumbai & Chennai, surveying and watching of samples referred by Deputy Drugs Controller (India), West Zone.

➢ New drugs and formulations are also being referred by the Drugs Controller General (India) for analysis.

The laboratory is notified as Appellate Laboratory for Copper T Intra-Uterine Contraceptive Device and Tubal Rings under the Drugs and Cosmetics Rules, (Medical Stores) & Regional Directors of Department of Family Welfare and procurement and field samples of Oral Contraceptive Pills, Copper T and Tubal Rings referred by the Dept. of Family Welfare.

Regional Drugs Testing Laboratory (RDTL), Guwahati

The Regional Drugs Testing Laboratory was set up at Guwahati in 2002 under the Drug & Cosmetics Act 1940. The laboratory was set up for the entire North Eastern State including Sikkim and is housed in its building at Guwahati. It is one of the five National Laboratory of the Govt of India for quality control of drugs and cosmetics. It has been functioning under the administrative control of the Drugs Controller General of India. It also works as a sub ordinate office under the Directorate General of Health Services, Ministry of Health & Family Welfare.

Regional Drugs Testing Laboratory (RDTL), Chandigarh

Regional Drugs testing laboratory, Chandigarh, has been functioning since November 2007. Presently, the laboratory is testing the drugs at an average of 50 samples per month, primarily to cater to the requirements of CDSCO (North Zone). The laboratory is in the process of up-gradation of its infrastructure and workers to increase its testing capacity.

Central Drugs Institute (CRI), Kasauli

Central Drugs Laboratory at Central Research Institute (CRI) Kasauli is a central laboratory and has been engaged in the testing of vaccines. It is a notified laboratory under the Drugs and Cosmetics Act, 1940 to function as a Central Drugs Laboratory for testing of the following drugs or classes of drugs;

● Sera

● Solution of serum proteins intended for injection

● Vaccines

- Toxins
- Antigens
- Anti-toxins
- Sterilized surgical ligature and sterilized surgical sutures
- Bacteriophages, including the Oral Polio vaccine.

Major Functions of CDSCO

The major functions of CDSCO are as

- Regulatory control over the import of drugs, approval of new drugs and clinical trials,
- To conduct meetings of the Drugs Consultative Committee (DCC) and Drugs Technical Advisory Board (DTAB),
- (the approval of certain licenses as Central Licence Approving Authority.

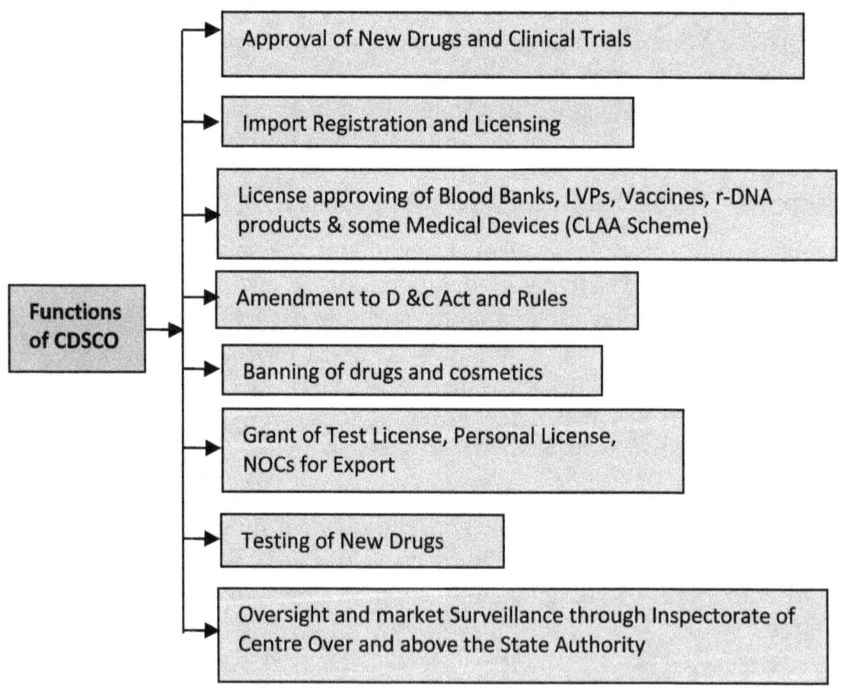

Fig. 5.3 Function of CDSCO

Indian Pharmacopoeial Commission (IPC)

The Indian Pharmacopoeial Commission has been registered as a Society under the provisions of the Societies Registration Act, 1860(Act No. 21 of 1860) for the registration of Literary, Scientific and Charitable Societies, on December 09, 2004.

Indian Pharmacopeia Commission is an autonomous institution under the Ministry of Health & Family Welfare, Govt. of India. It is dedicated to setting up standards for drugs, pharmaceuticals, healthcare devices/ technologies, etc. It also provides Reference Substances and Training. Its functions are

- ➤ To develop comprehensive monographs for the inclusion of the substances in the Indian Pharmacopoeia. The substance may be an active pharmaceutical ingredient (drug), excipients, and dosage forms and medical devices and to keep them updated revision regularly.

- ➤ To accord priority to monographs of drugs included in the national Essential Drugs List and their dosage forms.

- ➤ To prepare monographs for products that have normally been in the market for 2 years or more, except for certain special categories of new drugs like antiretrovirals, antituberculosis and anticancer drugs and their formulations introduced more recently, which may be accorded priority attention.

- ➤ To give special attention to the methods of manufacture used by the indigenous industry in selecting the pharmacopoeial tests for monitoring the toxic impurities of the concerned drug.

- ➤ To record the different levels of sophistication in analytical testing/instrumentation available while framing the monographs.

- ➤ To accelerate the process of preparation, certification, and distribution of IP Reference Substances, including the related substances, impurities, and degradation products required.

- ➤ To collaborate with pharmacopeia like the Ph Eur, BP, USP, JP, and International Pharmacopoeia to harmonize with global standards.

- ➤ To organize educational programs and research activities for spreading and establishing awareness on the need and scope of quality standards for drugs and related articles/materials.

State Licensing Authority

Every state Government constitutes the Directorate of Drug control administration as an enforcing department.

Organization

The organization for state Drugs Control department i.e., state licensing authority is as follows;

Thus, the state drug control department is administratively controlled by the Ministry of Health Sciences of the state. The State Government sanctions the posts and appoints the eligible candidates in various categories. The Central Drugs Standard Control Organization (CDSCO) has been discharging the functions

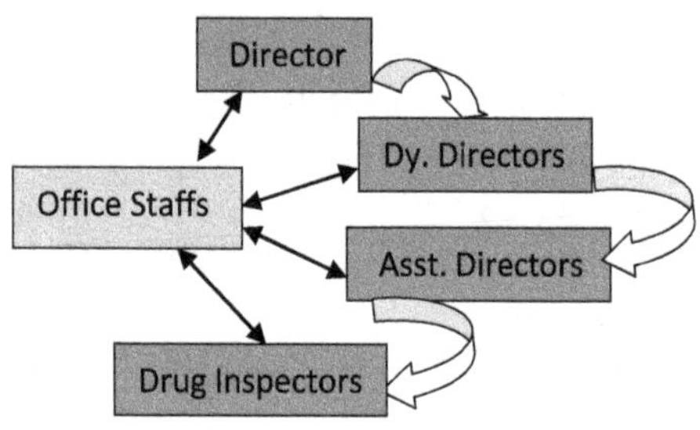

Fig. 5.4 Organization of State Drug Control Administration

assigned to the Central Government under the Drugs and Cosmetics Act and framing the rules and regulations for the manufacture, sale, export, import of drugs in the country.

The State Drug Control Department is responsible for the implementation of the Drugs & Cosmetic Act, 1940, and the rules there under.

Responsibility

The responsibilities of the state drug control department are mentioned below;

➢ Issue of forms for manufacture and sale of the drug and/or its product

➢ Verification of application form

➢ Inspection of the premises

➢ Grant of licenses

➢ Renewal of licenses

➢ Inspections at medical stores and

➢ Renewal of licenses.

The most important duty of the State Drug Control department is to ensure the supply of quality drugs at the price fixed by GOI (NPPA) to the people of the state. Therefore, the State Drug Control department is an enforcing wing of CDSCO. It also controls the drug testing laboratory of the state. In some states, the State Drug Control department monitors the education in pharmacy, particularly at the diploma level. Education in the state is controlled by the state Pharmacy councils who conduct the examination.

Enforcement Wing

The Enforcement Wing of the state deals with licensing of Drugs/ Cosmetics/Blood banks/Medical Devices manufacturing establishments, sales establishments, inspections of manufacturing and sales establishments and drawl of samples for test and analysis, investigation of complaints, launching of prosecution against the defaulters, etc. The department performs the work in a two-tier system as follows;

➢ Head office

➢ Circle office

➢ The levels of officers are among

➢ Drugs Controller

➢ Add. Drugs Controller

➢ Deputy Drugs Controller

➢ Assistant Drugs Controller

➢ Drugs Inspector

As per the Drugs & Cosmetics Act, 1940 and Rules, there under there are three functionaries;

- Inspector: The Central Government/State Government by notification in the Official Gazette appoints Drugs Inspectors under Section 21 of Drugs and Cosmetics Act, 1940 for areas assigned to them.

- Licensing Authority: Drugs Controller is the Licensing Authority for manufacture/sales establishments. However, with the approval of the state government, he delegates the powers of the Licensing Authority regarding sales establishments to Asst Drugs controllers working at Circle offices.

- Controlling Authority: Drugs Controller is the controlling authority. All inspectors are officially subordinate to Controlling Authority.

Functions of the Head Office

- The overall administration of the entire organization

- Grant/renewal of Manufacturing licenses for Drugs & Cosmetics, Approval of laboratories, Blood Banks and Blood components, Blood Storage Centres,

- Approval to Recognized Medical Institutions for transport, possession, and use of Oral Morphine for palliative care in cancer patients.

- Inspections of manufacturing units of Drugs, Cosmetics, Approved laboratories, Blood banks, Blood Storage Centres, and medical devices.

- The investigation of complaints.

- Drawl of samples from manufacturing establishments

- Issue of certificates such as GMP (Good Manufacturing Practices) certificate, Free Sale Certificate, Tender purpose Certificate, Export Registration Certificate, etc.,

- The overall administration of the entire organization.

- Three deputy drug controllers are placed at headquarters to monitor the administration and functioning of officers working in 6 circles of the Bangalore urban district.

Functions of the Regional Offices

Based on population, the State Drugs Controller divides the state into different regions and appoint officers of different levels. These offices are called regional offices. Regional offices may be headed by the Deputy Drugs Controller, under whom Asst. Drugs Controller and Drug Inspectors are posted. They remain at the overall incharge of their respective regional offices.

Generally, the functions of the regional offices are

- The overall administration of the entire jurisdiction assigned

- Blood Bank/Blood Storage Centres Inspections

- Collection and dissemination of information gathered during intelligence work

- Inspections of sales/manufacturing establishments whenever required

- Drawl of samples from sales premises/Drugs stores attached to Govt. hospital for test and analysis

- Investigation of complaints

- Any other work assigned by higher-ups.

One Drug Inspector is attached to the regional offices of the Deputy Drugs Controllers as Intelligence Officer.

Intelligence Wing

The main objective of this wing is to search and expose Spurious/Adulterated/Misbranded Drugs, Cosmetics and to detect unlicensed dealers/manufacturers. Usually, such a wing is headed by the Additional Drug Controller.

State Intelligence Branchs

Such an office or branch is also constituted by the State Drugs Controller and required several Drug Inspectors along with Dy. / Asst. Drugs Controller is deputed in the office.

Regional Intelligence

One Drug Inspector is attached to each regional office of the Deputy Drugs Controllers as Intelligence Officer having jurisdiction over circles.

Their functions are

- Collection and dissemination of information gathered during intelligence work
- Blood Bank/Blood Storage Centres Inspections
- Inspections of sales/manufacturing establishments whenever required
- Drawls of samples from sales/manufacturing premises/drug stores attached to Govt. / Pvt hospitals for testing and analysis
- Investigation of complaints
- Institution of prosecution and follow up
- Any other work assigned by higher-ups.

Under each regional office, there are various numbers of circles and in each circle, there may be one Circle office. Usually, each district is provided with one circle office. Depending on the requirement, the required number of officers and staff are provided.

Circle Offices

The powers of Licensing Authority have been given to the Drugs Controller under the Drugs and Cosmetics Rules, 1945 with respect to the jurisdiction assigned to each office.

The functions of a circle office include

- The overall administration of the entire jurisdiction assigned

- Grant/Renewal of sales licenses

- Inspections of sales/manufacturing (wherever required) establishments, Government/Private hospital drug stores, Blood Banks & Blood Storage centers

- Drawl of samples from sales/manufacturing (wherever required) establishments/Drugs Stores attached to Govt./private hospital for test and analysis

- Investigation of complaints

- Institution of prosecution against the defaulters & follow up

- Any other work assigned by a higher officer.

Common Technical Document (CTD)

The CTD is only a format for the submission of information to CDSCO. There is no guideline other than CTD, and it covers all necessary information. The safety and efficacy of a drug product, when used by human beings, is most essential. This should be established before its approval by the CDSCO for importing to or manufacturing in this country. As per, the Rules 122A, 122B, 122D, and Appendix I, IA, and VI of schedule Y of the Drugs and Cosmetic Rules, the applicant should apply for approval of the new drug. The application should contain substantial data support. Thus, a common format of submission as per the guidelines of the International Conference on Harmonization (ICH) has been introduced to meet all requirements. This is known as Common Technical Document (CTD). The guidelines were initially developed for Japan, European Union, and the USA. Since 2010 most of the countries have adopted the CTD format. Based on, (1) the ICH Harmonised Tripartite Guideline on "Organisation of the Common Technical Document for the Registration of Pharmaceuticals for Human Use" and (2) the Drugs & Cosmetics Act 1940 and Rules made thereunder, the CDSCO has developed this format for the registration of pharmaceutical products. A duly filled-in document should be submitted in one hard copy and three soft copies (CD). The CTD comprises 5 modules show in figure 5.5. The modules 1, 2, 3, 4, and 5 are briefly described below.

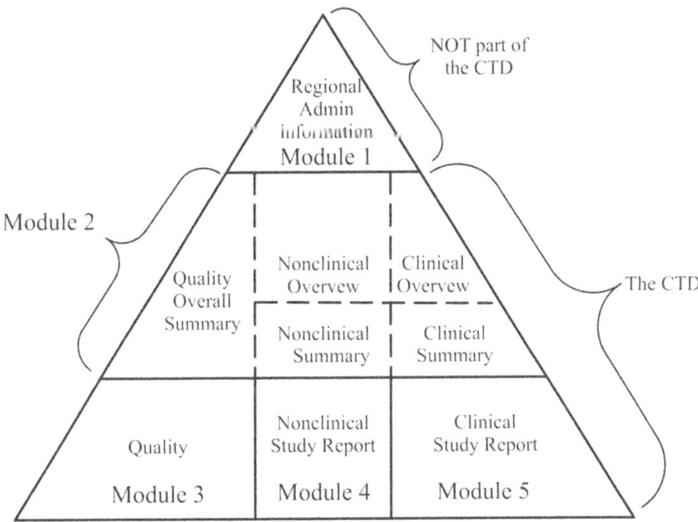

Fig. 5.5 Common technical document

Module 1: General Information

This module should contain documents specific to India—for example, Form 44, Treasury challan fee, or the proposed label for use in India.

Module 2: Quality Overall Summary

Introduction: This should include the proprietary name, non-proprietary name, or common name of the drug substance, company name, dosage form(s), strength(s), route of administration, and proposed indication(s).

Module 3: Quality

Information on quality should be presented in a structured format according to all the subsections of the quality guidelines. A. Drug substance: structure, general properties, the description of the manufacturing process and process controls, Control of critical steps, process validation, Impurities, Quality control of the drug substance, Reference standards, container closure system, Stability Data. B. Drug Product: Description and composition of drug product, Pharmaceutical development, Physicochemical and biological properties, Excipients, Microbiological attributes, Batch formula, Analytical Procedures, Control of excipients.

Module 4: Nonclinical Study Reports

The nonclinical study reports should be presented. For Generic drugs complete non-clinical study is not required to be shown as the safety data are already provided during the approval of the branded drugs.

Module 5: Clinical Study Reports

Human studies reports and related information should be presented.

Certificate of Pharmaceutical Product (COPP)

The United Nations (UN) was formed in 1945. One of their goals was to set up a global health organization. Currently, the United Nations Organization has 193 member states. The UN created the World Health Organization (WHO) whose constitution came into force on 7th April 1948. Today the WHO is the authority to direct and coordinate the functions related to health within the UN system. The responsibility of the WHO includes;

- Discussions on global health topics in general,

- Research on health issues,

- Providing norms and standards for research and health industry, and

- Creation of guidelines available for all countries worldwide.

- Providing technical support to countries worldwide and further monitoring of health trends, including their assessment.

The WHO harmonizes the responsibility of all countries to make pharmaceutical products and essential care available to all humans worldwide and further to provide a defense against transnational threats like epidemic diseases. Using the structure of the WHO A single country or a single resource cannot provide so much support on the different areas of work represented by the WHO. The WHO with many participating countries, has utilized and accumulated several resources of knowledge. The vision of the WHO is that people everywhere have access to the essential medicines and health products they need and further the medicines and health products are safe, effective and of assured quality; and that medicines are prescribed and used rationally.

In the health sector, different guidelines and recommendations have been issued by the WHO. The WHO provides guidelines and strengthens its medicine strategy so that people across the world can access medicine. One of these strategies is the provision of extending scientific or medicinal support and ensuring the quality and safety of medicinal products throughout the world. Guidelines created by the WHO are approved by the Guidelines Committee (GRC). The GRC includes content experts, methodologists, target users, policymakers with reviewers of gender and geographical balance.

One of the guidelines created by WHO within this area of work is the WHO Certification Scheme on the quality of pharmaceutical products as a voluntary and non-binding agreement between WHO and their Member States. This has been done to provide a comprehensive system of quality assurance and a reliable system of licensing. According to the WHO, this certificate can be said a Certificate in conformity.

The objectives of the WHO Certification Scheme are;

➢ To provide information on the Quality, Safety, and Efficacy (QSE) of imported finished medicinal products,

➢ To facilitate the appropriate use of the guideline, and

➢ To provide a reliable system of licensing with independently controlled quality according to the accepted norms.

The CPP should be considered the condition for approval, not just for submission. According to the WHO Certification Scheme, the content of certificates can easily be transformed into national templates to provide the information by the country for a specific use. But, the scope of this scheme should not be expanded by supplementing the content of the certificates.

The World Health Organization (WHO) issues the certificate of pharmaceutical products (CPP or COPP) in the prescribed format and recommends the pharmaceutical products and the applicants to the exporting country. The certificate is issued for a single product because manufacturing conditions and approved information for different pharmaceutical products and strengths can vary.

The certificate of a pharmaceutical product is required by the importing country when the concerned product is intended to be registered (licensing, authorization) or renewed the registration (prolongation) for commercialization or distribution in that country. With the issuance of such CPP, the WHO also recommends the national authorities to

➢ To verify and confirm the analytical methods in their national laboratory,

➢ To review, if necessary,

➢ To adapt product information as per the local labeling requirements, and

➢ To assess bioequivalence and stability data, if necessary.

The regulatory practices in some importing countries may be different. However, in addition to CPP, the application dossiers are assessed for drug registration. There may be different levels and complexity of requirements to satisfy full assurance of the appropriate quality of drugs.
It has already been mentioned that the WHO Certification Scheme is a voluntary agreement under which countries can apply for participation. WHO Member States (WHO MSs) or regional organizations can notify the Director-General of the WHO and confirm that they want to participate in the Scheme. The confirmation should be made through a written notification. The name of the Competent Authority (CA) must be named in writing, which will issue the relevant certificate. MSs, participate in the WHO Certification Scheme, should have the administrative capacity for issuing the requested CPPs. Of course, they must provide a registration system for medicinal products assuring quality, safety, and efficacy (QSE).

The certification scheme is not mandatory to participation. It may be possible for the countries that have limited regulatory capacity in the healthcare and medicinal sector. Because obtaining a declaration from the manufacturer of the exporting country of a QSE pharmaceutical product needs good infrastructure and this needs to be confirmed by the competent authority (CA) of the country of origin[4] (CoO). The WHO provides a list of addresses of competent national regulatory authorities currently participating in the Scheme.

Information to be submitted to COPP

No. of certificate

Exporting (certifying country):

Importing (requesting country):

1.   Name and dosage form of the product:

1.1.   Active ingredient(s) and amount(s) per unit dose:

For complete composition, including excipients:

1.2   Is this product licensed to be placed on the market for use in the exporting country? (yes/no)

1.3   Is this product actually on the market in the exporting country?

If the answer to 1.2. Yes, continues with section 2A and omit section 2B.

If the answer to 1.2 is no, omit section 2A and continue with section 2B

2. A.1.   Number of product licenses and date of issue

2. A.2.   Product license holder (name and address)

2. A.3.   Status of the product license holder

2. A.3.1. For categories b and c, the name and address of the manufacturer producing the dosage form is

2. A.4.   Is a summary basis for approval appended? (yes/no)

2. A.5.   Is the attached, officially approved product information complete and consonant with the license? (yes/no/not provided)

2. A.6.   An applicant for a certificate, if different from the license holder (name and address)

2. B.1.   An applicant for the certificate (name and address)

2. B.2.   Status of applicant

2. B.2.1. For categories (b) and (c) the name and address of the manufacturer producing the   dosage form

2. B.3.   Why is marketing authorization lacking? (not required/not requested/under consideration / refused)

2. B.4.   Remarks

3.   Does the certifying authority arrange a periodic inspection of the manufacturing plant at which the dosage form is produced? (yes/no/not applicable); if not or not applicable, proceed to question.

3.1.   Periodicity of routine inspections (years)

3.2.   Has the manufacturing of this type of dosage form been inspected? (yes/no)

3.3   Do the facilities and operations conform to the GMP as recommended by the World Health Organization?

4.   Does the information submitted by the applicant satisfy the certifying authority on all aspects of the manufacture of the product: (yes/no)

If no, explain

Address of certifying authority:

Telephone:

Fax:

Name of authorized person:

Signature

Stamp and date

(WHO, World Health Organization - Medicines, 2013)

Explanatory notes

1. This certificate, which is in the format recommended by the WHO, establishes the status of the pharmaceutical product and the applicant for the certificate in the exporting country. This is used only for a single product only. The manufacturing arrangements and approved information for different dosage forms and different strengths can vary; hence, separate application is necessary.

2. Use, whenever possible, International Non-proprietary Names (INNs) or national non-proprietary names.

3. The formula (complete composition) of the dosage form should be mentioned on the certificate or be appended.

4. Details of quantitative composition are preferred, but their provision is subject to the agreement of the product-license holder.

5. When applicable, append details of any restriction applied to the sale, distribution, or administration of the product that is specified in the product license.

6. Sections 2A and 2B are mutually exclusive.

7. Indicate, when applicable, whether the license is provisional, or the product has not yet been approved.

8. Specify whether the person is responsible for placing the product in the market:

    a) Manufactures the dosage form;

    b) Packages and/or labels a dosage form manufactured by an independent company; or

    c) Is involved in none of the above.

9. This information can only be provided with the consent of the product license holder or, in the case of non-registered products, the applicant. Non-completion of this section indicates that the party concerned has not agreed to the inclusion of this information. Note that information concerning the site of production is part of the product license. If the production site is changed, the license has to be updated or it is no longer valid.

10. This refers to the document, prepared by some national regulatory authorities, that summarizes the technical basis on which the product has been licensed.

11. This refers to the product information approved by the competent national regulatory authority, such as Summary Product Characteristics (SPC)

12. Under this circumstance, permission to issue the certificate is required from the product-license holder. This permission has to be provided to the authority by the applicant.

13. Please indicate the reason that the applicant has provided for not requesting registration.

    a) The product has been developed exclusively for treating conditions — particularly tropical diseases — not endemic in the country of export;

    b) The product was reformulated to improve its stability under tropical conditions;

    c) The product was reformulated to exclude excipients unapproved for use in pharmaceutical products in the country of import;

    d) The product was reformulated to meet a different maximum dosage limit for an active ingredient;

    e) Any other reason, please specify.

14. Not applicable means the manufacture is occurring in a country other than that issuing the product certificate and inspection is conducted under the aegis of the country of manufacture.

15. The requirements for good practices in the manufacture and quality control of drugs referred to in the certificate are those included in the thirty-second report of the Expert Committee on Specifications for Pharmaceutical Preparations, WHO Technical Report Series No. 823, 1992,

    Annex 1. Recommendations specifically applicable to biological products have been developed by the WHO Expert Committee on Biological Standardization.

16. This section is to be completed when the product license holder or applicant conforms to status (b) or (c) as described in note 8 above. This is particularly important when foreign contractors are involved in the manufacturing of the product. In these circumstances, the applicant should supply the certifying authority with information to identify the contracting parties responsible for each stage of manufacture of the finished dosage form, and the extent and nature of any controls exercised over each of these parties.

Issuing a Certificate

As the CPP is classified as a confidential document, it can only be issued by the Competent Authority (CA). This is done only when the applicant applies for the certificate and permits the requesting health authority (HA) to provide a CPP for a specific finished medicinal product. The marketing authorization holder (MAH) may be different from the applicant. Usually, the applicant forwards the CPP, issued by the CA, to the local affiliate in the requesting country. The applicant also requests the relevant HA to send the CPP, along with the new drug application (NDA). Most competent authorities can issue a bilingual CPP. A CPP issued by the European Medicines Agency (EMA) is usually in English, on the request of the applicant it can also be issued in Spanish or French language. But, one CPP is signed by the authority, another one is translated copy. Usually, the responsibility for providing a translation is considered by the applicant of the NDA/supplemental registration.

The CA can charge fees for issuing a CPP. By issuing the CPP, the CA confirms the authenticity of the certified data. The CPP is issued by the CA of the participating member state (MS) of the WHO. The certificate does not bear a WHO emblem; but, it is considered if the content is as per the format recommended by the WHO. According to the recommended WHO format, the GMP status will also be reflected in the CPP; this implies that the CA should prove that the applicant/MAH manufactures the pharmaceutical products according to GMP standards at the registered manufacturing site. Note that this information does not guarantee the GMP status of the manufacturing of the active substance. The certifying authority should evaluate the CPP if it contains adequate information regarding compliance with the GMP according to WHO recommendations. This is done when the manufacturing and packaging of the finished product occur in different production sites until the final release. Usually, the CPP contains the official stamp or seal of the CA and every single CPP is a true original. But, some HAs require that the CPP is additionally legalized by a notary, country court, and/or embassy. A CPP is usually issued for the requesting country by naming the country within the CPP as the importing country. Any additional information, other than the content of the CPP and not recommended by the WHO, is submitted along with the CPP, must be labeled as such. If forgery is suspected, an identical copy can be demanded by the importing country directly at the CA of the CoO. Every CPP bears an individual identifier or unique number, such as a certificate number, by the issuing CA and the authenticity can be traced easily. Only participating MSs of the WHO are eligible to issue CPPs for registered finished medicinal products through the Competent Authorities in their countries according to their responsibility. National authorities can issue CPPs for products having a national registration. Within the European Union (EU) and the European

Economic Area (EEA), it is possible to obtain registrations through the centralized procedure (CP) with the EMA as the CA. For these centrally registered finished medicinal products, the CPPs can only be issued by EMA instead of the local national Competent Authorities in the member states of the EEA. Note that a CPP is never directly issued by the WHO.

The CA doesn't need to be in the CoO/exporting country where the finished product is manufactured and released, and it issues a CPP. Initially, it was thought that the CPP can be issued by the CA of the exporting country/the CoO or any other country, which has performed a complete review of QSE according to ICH guidelines and as per the WHO Certification Scheme.

According to the WHO, other countries besides the CoO can also issue a CPP when they approve the registration with a review of the QES. Countries can, therefore, issue different types of CPPs to identify the status of the CPP. For example, the U.S. Food and Drug Administration (U.S. FDA) of the United States of America (USA), attaches different colored ribbons to the CPP, which specifies the content details. As per the US FDA, three different types of CPPs can be differentiated as mentioned below:

- A Red Ribbonshall be attached to all (regular) CPPs issued for authorized medicinal products legally marketable in the US as authorized by the U.S FDA.

- A *Blue Ribbon*will be affixed to CPPs issued only for the export of unapproved medicinal products which are not authorized by the U.S. FDA but which may be legally exported.

- A *Yellow Ribbon*will be affixed to CPPs issued to products that are manufactured at sites outside of the USA.

Apart from these three types of CPPs, there is another specific type of U.S. FDA CPP:

- The Pilot-CPPis issued by the U.S. FDA on the products which are neither manufactured in the U.S. nor exported from the U.S.; only when no other country has approved the relevant registered finished medicinal products worldwide. This was done not to delay further approval of the respective products.

The U.S. FDA confirms these different types of CPPs with slightly different contents, even though they are always reflected in the finished drug product. For example, CPP with the *Red Ribbon*: They confirm that the drug product described in the CPP is the same which is manufactured and approved in the U.S.; such products may be exported to other countries. They will provide proof of the registered product and the approved label as well.

This certification reflects the intended use of the CPP according to the WHO Certification Scheme. This CPP confirms that the quality and safety of the product have been approved within a "comprehensive system of quality assurance... on a reliable system of licensing and independent analysis of the finished product..." including manufacturing according to GMP.

If an MS participates in the WHO Certification Scheme and issues a CPP in favor of the finished medicinal product which is not manufactured in the MS issuing the CPP, the registration of the specific finished product might be attested including the GMP status of the product. The issuing CA can confirm that the relevant HA in the issuing country reviewed the QSE of the certified finished medicinal product, but the confirmation of the GMP status is based on the information provided. It is the responsibility of the CA to confirm physically.

The WHO Q&A approach to confirming the GMP and registration status of a finished medicinal product in the range of the CPP avoids a delay of availability of important medicines worldwide. This approach can reduce the scarcity and increase access for patients. Even if the Drug Product (DP) is not registered in the country, the CPP is issued by the country. However, the WHO does not recommend the need for multiple CPPs from different countries because they do not provide any additional benefit.

Requesting Certificates

Usually, an application is submitted by the pharmaceutical company to the eligible competent authority (CA) to issue CPP for commercial purposes. The CA of the country of origin (CoO) or any other country issues a CPP according to the WHO Certification Scheme in favor of a locally registered finished medicinal product. On receipt of the CPP, the marketing authorization holder (MAH) or the applicant will afterward send the CPP to the requesting HA in the importing country. This is usually done through the local MAH of the importing/recipient country. This process is also depicted below.

According to the information provided by the WHO, the certification scheme for the quality of pharmaceutical products is "applicable to finished dosage forms of pharmaceutical products intended for administration to human beings or food-producing animals."

The GMP status of the manufacturing site is documented in the CPP. This indicates that independent inspections are required to ensure whether all manufacturing operations and processes are carried out as per the defined norms. These inspections are carried out and licensed under the GMP, which can be documented with a certificate for a specific manufacturing site of medicinal product production.

As per the recommendations of the WHO, the Good Manufacture Practices and Good Quality Control of drugs are referred to as GMP. This is defined by the WHO initially in 1969 including 4 revisions throughout the years 1975, 1988, 1992, and 1997.

Member States are advised to adopt and apply the internationally recognized and respected standards laid in the WHO guidelines.

The summary of product characteristics (SmPC), patient information leaflet (PIL), and Labelling attached to the CPP is used to provide additional information for importing countries. This information is mentioned as the approved product information registration in the exporting country. This can be highly relevant during label updates in the existing registrations in the exporting and importing country. The importing country requested the CPP from the MAH of the CoO not only during Quality, Safety, and Efficacy (QSE) review but also to establish any changes on the wording of the labeling in the CoO

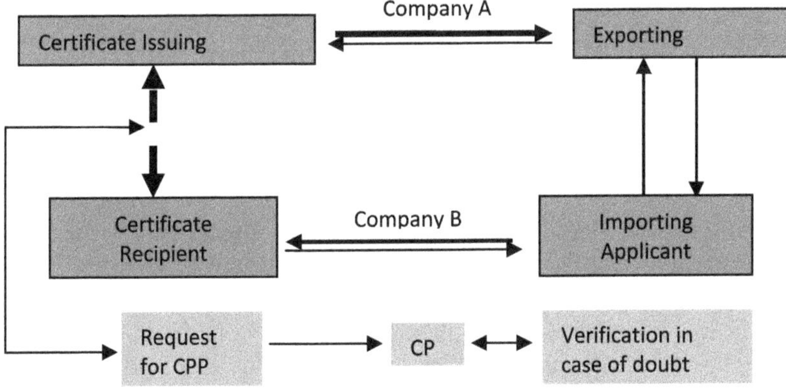

Fig. 5.6 Process of COPP application

Regulatory Requirements

The Import, manufacturing, sale, and distribution of the drug in India is regulated under the Drugs and Cosmetics Act 1940 and the Drugs and Cosmetic Rules 1945 made there under. At present, the bulk drug (Active Pharmaceutical Ingredients) and finished formulations are regulated under the said Act. Any substance falling within the definition of the drug (Section 3b of the Act) has to be registered before import into the country. Not only drugs but the manufacturing site needs to be registered for import. If the drugs, fall within the definition of New Drug (Rule 122 E of the Act), the new drug approval is pre-requisite for submission of application for Registration and/or import of the drug. The application for Registration and import can be made to the Licensing Authority under the Act; that is to the Drugs Controller General (I) at CDSCO, FDA Bhawan, Kotla Road, Near Bal Bhawan, New Delhi. The application should be submitted by the Local Authorized Agent of the foreign manufacturer having either manufacturing or sale License or by the foreign manufacturers having a wholesale License in the country.

Guidance

This guidance applies to those drugs manufactured outside India, and the import registration is to be issued (under Form 41) by the Central Drugs Standard Control Organization,(CDSCO) Directorate General of Health Services, Ministry of Health and Family Welfare, Government of India.

➢ An application shall be made to the Licensing Authority in Form 40, either by the manufacturer himself, having a valid wholesale license, for sale or distribution of drugs, or by his authorized agent in India either having a valid License to manufacture for sale of a drug or having a valid wholesale License for sale or distribution of drugs.

➢ Information to be provided in Form 40: the authorized signatory name, designation, department, along with the complete address of the company.

- *Authorized Signatory:* The authorized signatory is the person preferably the Director approved by the Board of Directors in case of a company or by the proprietor in the case of a proprietorship firm is considered. The application should contain an affidavit with respect to the authorized person or the Power of Attorney in the name of the authorized person.

  The Form should furnish the details of the Foreign Manufacturer's contact person, complete address of the manufacturing site with corporate office address, along with the telephone number, Fax number, and E-mail address.

a) The address of manufacturing premises shall be below: the responsible person at the manufacturing site (contact person in the manufacturing site) should give undertake of the document contents–

b) Regarding the import of more than one drug or a class of drugs manufactured by the same manufacturer, provided that drug or the classes of drugs are manufactured at one factory or more than one factory functioning conjointly as a single manufacturing unit.

c) In the respect of the drugs manufactured in two or more factories situated in different places, for the manufacturing of the same or different drugs the name and address of both the manufacturing sites should be mentioned, for example, if the tablets are manufactured at one location and packed at another location, name, and addresses of both the locations indicate the activity of each location.

- The Form shall contain the complete and correct names of the Drugs to be imported in India.

  a) The names of the drug(s) shall be written below:

  b) The brand name shall be mentioned clearly.

  c) Different packs, pack sizes, and/or different strengths of the same brand shall be mentioned.

    Importer's undertaking letter declaring that the information specified in Schedules D (I) and Schedule D (II), is provided by the original manufacturer.

    The registration Fees amount (Challan number and date) shall be mentioned on the Original TR 6 challan having the complete name and address of the applicant and details of the application to be enclosed.

- ➢ *Fee structure for Import Registration under Form 40*: Fees and Form(s) and the undertakings as per Schedule D(I) (for registration of the manufacturing premises) and Schedule D(II) (for registration of the drugs) is mentioned below:

  a) The applicant shall make a payment of 1500 USD (or its equivalent to Indian Currency), as a registration fee for the Manufacturing premises.

b) The applicant shall make a payment of 1000 USD (or its equivalent to Indian Currency), as a registration fee for a single drug and an additional fee of 1000 USD for each additional drug when the manufacturing site remains the same.

Fees shall be paid through a Challan in the Bank of Baroda, Kasturba Gandhi Marg, New Delhi 110 001, or any other Bank, as notified, from time to time by the authority.

➤ *Challan and Bank details*: Fees shall be paid through a Challan in the Bank of Baroda, Kasturba Gandhi Marg, New Delhi-110 001, or any other Bank, as notified, from time to time by the authority.

The pay order cheque to be in favor of Pay & Accounts Officer, DGHS, Nirman Bhavan, New Delhi-01

Challan means-the receipt of the Cash paid into the bank, which is attested by the bank with seal and date.

The fee is to be credited under the following details

a)   Head of Account

b)   104 – Fees and Fines under the drugs and cosmetic rules 1940. The conversion rate should be mentioned on the Challan.

• Electronic Payment: In case of any direct payment of fees by the manufacturer in the country of origin through ECS (Electronic Clearance System) from any bank in the country of origin to Bank of Baroda, Kasturba Gandhi Marg, New Delhi-110 001, through the Electronic Code of the Bank in the Head of Account, —104 – Fees and Fines‖ and Original receipt of the said transfer can be considered as an equivalent to bank Challan, subjected to approval by Bank of Baroda.

Applicant must pay 5000 USD (or its equivalent to Indian Currency) for Expenditure (Inspection fees + expenditure on inspection to be borne by company) as may be required for Inspection or Visit of manufacturing premises.

The applicant shall make the payment of testing fees directly to a testing laboratory approved by the central government in India or abroad, as required for examination testing and analysis of drugs.

The applicant has to pay a fee of 300 USD (or its equivalent to Indian Currency) for a copy of the registration certificate, if the original is defaced, damaged, or lost.

➤ *Registration time*: if the application is complete in all respects and information specified in D (I) and D (II) are in order, the licensing authority shall issue such Registration Certificates within 9 months from the date of receipt of application.

➤ In exceptional circumstances and for the reasons recorded in writing, the Registration Certificate may be issued within such an extended period not exceeding 3 months as the licensing authority may be deemed fit.

- To obtain, the registration certificate and to keep the validity of the registration certificate the applicant should give an undertaking for the compliance of the terms and conditions required.

- The data specified in Schedule D (I) and Schedule D (II) shall be enclosed along with the covering letter and Table of content.

Details are to be mentioned in the Covering Letter

➤ Information on the drugs to be imported

➤ Manufacturer information like address and contact details.

➤ Brief information about the application and List of Documents-

- Original Challan and the details of the Challan

- Form 40

- Schedule D(I) documents as mentioned by the drug(s) manufacturer (Module 1 of CTD format)

- Schedule D(II)-documents as mentioned by the drug(s) manufacturer (Module 2 to 5 of CTD format)

- Power of Attorney issued by the manufacturer

- Copy of Whole Sale License of applicant

- Copy of Authorization letter of Applicant

- An undertaking shall be submitted by the proprietor of the firm in case of proprietorship firm and in the case of Private Limited Company by the board of Directors.

Registration of bulk drug(s) and finished formulation(s) in India (For Common Submission)

For the Import and Registration of the bulk drug(s) and finished product(s) in India the following documents are to be submitted to the CDSCO in the following manner and order.

Applicants are requested to apply to 3 or more different files as follows:

1. The covering letter is an important part of the application and should specify the aim of the application

   - whether the application for the registration of the manufacturing site is being submitted for the first time,

   - whether the application is for re-registration/renewal

   - whether the application is for endorsing additional products to an existing Registration Certificate,

   The covering letter should be duly signed and stamped by the authorized signatory, indicating the name and designation of the authorized signatory along with the name and address of the firm. Any exemption to the submission requirement is specified in the covering letter on the firm/company letterhead and justified in the submissions.

   The list of documents submitted (Index with page nos.) As well as any other important and relevant information may be provided in the covering letter.

   A Resolution shall be submitted by the proprietor of the firm in case of a proprietorship firm and by the board of Directors in case of Private Limited Company/firm.

2. The authorization letter should be in original which is issued by the Director/Company Secretary/Partner of the Indian Agent firm disclosing the name and designation of the person authorized to sign (along with the name and address of the firm) legal documents such as Form 40, Power of Attorney, etc. On behalf of the company should be submitted at the time of submission of the application for registration (Rule122A). It should have a validity period as per the company's policies. Duly self-attested photocopies of the Authorization letter may be submitted at the time of submission of subsequent applications.

3. A duly filled and signed Form 40 as per the proforma prescribed in the Drugs & Cosmetics Rules, signed & stamped by the (Local Authorized Agent/manufacturer) along with name & designation and date is to be submitted. Form 40 should be signed by the (Local Authorized Agent or manufacturer and should have a valid sale or manufacturing License in India.

4. In case of any direct payment of a fee by the manufacturer in the country of origin, the fee shall be paid through the Electronic Clearance System (ECS) from any bank in the Country of Origin to the Bank of Baroda, Kasturba Gandhi Marg, New Delhi, through the

electronic code of the bank in the head of Account stated above and the original receipt of the said transfer shall be treated as an equivalent to the Bank Challan (TR 6 Challan), subject to the approval by the Bank of Baroda that they have received the payment.

5. The authorization by a manufacturer to his agent in India shall be documented by a Power of Attorney. It must be executed and authenticated either India a First Class Magistrate, or in the country of origin before such an equivalent authority. The certificate is to be attested by the Indian Embassy of the said country, and the original of the same shall be furnished along with the application for Registration Certificate. Apostille Power of Attorney from Hague convention member countries is also acceptable. The authorized agent will be responsible for the manufacturer's business activity, in India.

While submitting the Power of Attorney, the following points should be kept in mind:

➢ It should be co-jointly signed and stamped by the manufacturer and the Indian Agent, indicating the name and designation of the authorized signatories (along with the name and address of the firm).

➢ It should list the names of all proposed drugs if possible, along with their specific indication and/or intended use.

Further, the names of the proposed drug should correlate with those mentioned in Form 40, Free Sale Certificate, or Certificate of pharmaceutical product (COPP) as per the WHO-GMP certification scheme. The names and addresses of the manufacturer (Contract manufacturer name from different sources) as well as the Indian Agent stated in the Power of Attorney should correlate with Form 40. Multiple sites are to be mentioned in tabular form. It should be valid for the period of said Registration Certificate. This implies that a fresh POA is to be submitted at the time of revalidation of the Registration Certificate.

6. A duly attested/notarized (in India) and valid copy of Wholesale License for sale or distribution of drugs under Drugs and Cosmetics Rules in Forms 20B & 21Bor its renewal in Form 21Cissued to the manufacturer (subsidiary office/representative of the parent company) or its agent by the State Licensing Authority in India is to be submitted. If the agent is a manufacturer, a duly attested/notarized in India and a valid copy of manufacturing License issued by the State licensing Authority is to be submitted.

7A) Schedule D(I) undertaking as per the proforma prescribed in the Drugs & Cosmetics Act &Rules, signed & stamped by the manufacturer/Authorized agent indicating the name and designation of the authorized signatory is required to be submitted.

B) The Plant Master Fileasperin the prescribed format.

Schedule D (II) requirements are

8A) As per the second schedule of the Act, thestandards of the drugare to be submitted with a sample of imported drugs or the drug(s) manufactured for sale/sold stocked or exhibited for sale or distributed in the country. If the drug is in IP, it must meet the standards of identity, purity, and strength otherwise it must meet the standards of USP, BP, or EP.

B) True copy of the label as per Rule 96 of the Act is to be submitted. If the drug is on the IP label claim must be as per the IP.

C) In case of registration of Bulk (Active Pharmaceutical Ingredient) drugs of three consecutive batches are to be submitted to the designated laboratory for testing for which a fee is to be paid by the applicant to the Laboratory as per their norms. All the test reports are to be submitted.

The applicant should enclose adequate samples for reanalysis from each of the three consecutive batches along with specifications, the method of analyses, the certificate of analysis (COA) in the laboratory, impurity Standards, marker compounds, Reference Standard with its COA where ever applicable.

9) Duly notarized/Apostilled/Attested (by Indian Embassy the country of origin) and valid copy of Free Sale Certificate/Certificate to Foreign Government/Certificate of Marketability/for each drug issued by the National Drug Regulatory Authority of the country of origin is to be submitted. Free Sale Certificate should state that the proposed drug is freely sold in the country of origin and can be legally exported.

10) Duly notarized/Apostilled/Attested (by Indian Embassy the country of origin) and valid copy of GMP Certificate of WHO guideline or Certificate of Pharmaceutical Product (COPP) as per WHO GMP Certification Scheme/Product Registration Certificate issued by NRA and or proof of DMF approval by NRA and/or CEP (EDQM certificate) for each drug issued by the National Drug Regulatory Authority of the country of origin is to be submitted.

11)   Duly notarized/Apostilled/Attested (by Indian Embassy the country of origin) and valid copy of the Manufacturing License and or Market Authorization Certificatein respect of applied drugs issued by the National Drug Regulatory Authority of the country of origin) is to be submitted. Free sales certificates from other countries if available are also be submitted.

12)   Duly notarized/Apostilled/Attested (by Indian Embassy in the country of origin) and valid copy of Product Registration Certificate wherever applicable regarding the foreign manufacturing sites)

FORM 40

(See rule 24-A)

Application for issue of Registration Certificate for import of drugs into India under the Drugs and Cosmetics Rules 1945

I/We*____

______________________________ (Name and full address) hereby apply for the grant of Registration Certificate for the manufacturer, M/s.

______________________(full address with telephone, fax and E-mail address of the foreign manufacturer) for his premises, and manufacturing drugs meant for import into India.

1. Names of drugs for registration.

2. I/We enclose herewith the information and undertakings specified in Schedule D

(I) and Schedule D (II) duly signed by the manufacturer for the grant of Registration Certificate for the premises stated below.

3. A fee of............................. for registration of premises, the particulars of which are given below, of the manufacturer has been credited to the Government under the Head of Account "0210-Medical and Public Health, 04-Public Health, 104-Fees and Fines‖ under the Drugs and Cosmetics Rules, 1945-Central *vide* Challan No._________ Dated____________________(attached in original).

4. A fee of.......................... For the registration of the drugs for import as specified in Serial No. 2 above has been credited to the Government under the Head of Account "0210-Medical and Public Health, 04-Public Health, 104-Fees and Fines" under the Drugs and Cosmetics Rules, 1945-Central vide Challan No._______, Dated________ (attached in original).

5. Particulars of premises to be registered where manufacture is carried on the following:

Address(s) ____________

Telephone No.____________ Fax____________________

E-mail________________

I/We* undertake to comply with all terms and conditions required to obtain a Registration Certificate and to keep it valid during its validity period.

Place: ____________

Date: ____________

*Signature*____________

*Name*____________

*Designation*________

Seal/Stamp of the manufacturer or his authorized agent in India.

(**Note**: In case the applicant is an authorized agent of the manufacturer in India, the Power of Attorney is to be enclosed). *Delete whichever is not applicable.*

Approval Procedures for New Drugs in India

When a company in India wants to manufacture/import a new drug, it has to apply to seek permission from the licensing authority (DCGI) by filling in Form 44 and also submitting the data as given in Schedule Y of Drugs and Cosmetics Act 1940 and Rules 1945. Application for the importation and manufacture of new drugs or to undertake clinical trial is made on Form 44 to the Central Drugs Standard Control Organization (CDSCO) together with the appropriate fees.

The approval for import permission is given in Form 45 or 45A, clinical trial on Form 46 and/or 46A, and new bulk drug substance on Form 54a.

Important forms of the Drug and Cosmetic Act

o Form 44 Application for a grant of permission to import or manufacture a new drug or to undertake a clinical trial

o Form 12 Application for license to import drugs for the purposes of examination, test, or analysis o Form 11 License to import drugs for examination, test, or analysis (Validity is one year)

o Form 4 Issue of import certificate (Validity is six months)

o Form 1 Application for the issue of an Import certificate for the import of narcotic drugs and psychotropic substances

Stages of approval

- Submission of Clinical Trial application to evaluate safety and efficacy.

- Requirements for permission of new drugs approval

- Post-approval changes in biological products: quality, safety, and efficacy documents.

- Preparation of the quality information for drug submission for new drug approval.

In India, schedule Y of the Drug & Cosmetic Act 1940 and Rules 1945 there under regulates the import and manufacture of new drugs. Schedule Y specifies the requirements and guidelines for permission to manufacture/ import new drugs or to undertake clinical trials.

The Rules that govern the import, manufacture, etc. of new drugs are;

- Rule 122A regulates the permission to import the new drug,

- Rule 122B regulates the manufacture of new drugs,

- Rule 122D regulates the permission to import/manufacture of FDC,

- Rule 122DB describes the conditions of suspension/cancellation of permission,

- Rule 122E defines the new drug.

- Rule 122DA regulates the permission to conduct clinical trials

- Rule 122DAA defines the clinical trial

- Rule 122DAB speaks about the compensation in case of injury or death during the clinical trial

- Rule 122DAC speaks the conditions of clinical trial permission and inspection

- Rule 122DD regulates the registration of the Ethics Committee.
  As per the Rule 122E, a new drug is defined as follows;

➤ Any new substance proclaimed for therapeutic use

➤ An already approved drug with modified or new (therapeutic claims, indications, dosage forms, or routes of administration

➤ FDCs of drugs

➤ All vaccines and recombinant r-DNA derived products

A new drug shall continue to be considered a new drug for four years

The information required for the import/manufacture of new drugs in India is;

➤ Chemical and Pharmaceutical information

➤ Animal Pharmacology

➤ Animal Toxicology

➤ Phase I, II, III Clinical Trials

➤ Regulatory Status in another country

➤ COPP/FSC (in case of import)

➤ Label, Prescribing information

The new drug already approved in other countries is permitted in India when adequate numbers of Indians have participated in phase III clinical trials in India as a part of global clinical trials. Different committees for evaluating applications are

- IND Committee – Evaluation of Investigational New Drugs (new molecules discovered in India). Chaired by DG, ICMR & Secretary, Department of Health Research.

- Subject Expert Committees (SEC) – 25 panels of about 350 various medical experts for evaluating applications of clinical trials and new drug approvals except IND,

- Technical Committee (TC) – Separate committee of experts chaired by the Director General of Health Services to review proposals Clinical trials and cases of CT waiver for new drug approval.

- Apex Committee: review recommendations of the Technical Committee in cases of clinical trials of NCEs, including IND.

Evaluation Steps of application for import/manufacture of a new drug in India is done as follows:

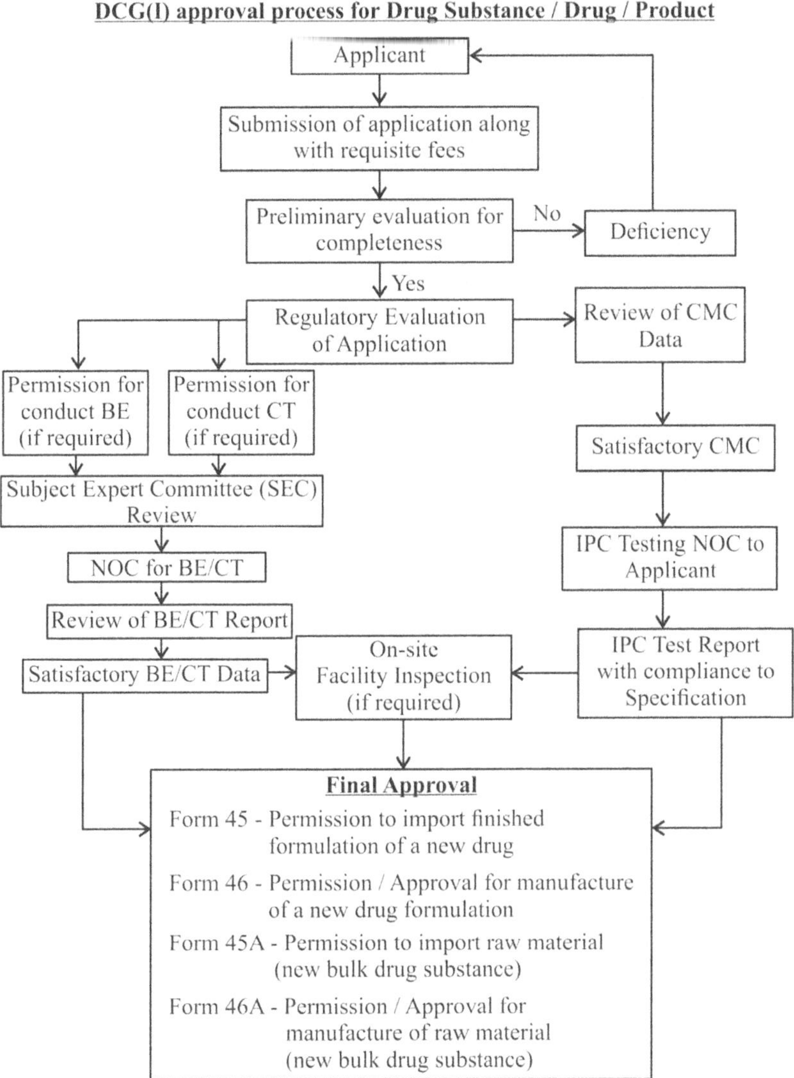

Fig. 5.7 Drug Approval Process in India for Drug Substance/Drug Product

Clinical Trial in India

Under the following circumstances the clinical trials are conducted in India;

➢ For New Drug substances discovered in India, clinical trials must be conducted right from Phase I with Indian patients.

➢ For new drugs approved outside India, Phase III studies are now being conducted to demonstrate the required efficacy and safety of the drug in Indian patients.

Recent initiatives in approval of the new drug in India;

➢ Online submission and processing of CT and new drug applications through SUGAM Portal

➢ Interactive workshop for the SEC members and reviewers of CDSCO for streamlining the new drug evaluation process in collaboration with USFDA and ICMR

➢ Guidelines for SECs for the review of clinical trial and new drug applications

➢ A comprehensive review of existing regulatory provisions and preparation of new rules to make them contemporary to meet the aspirations of the stakeholders are underway. The new rules will have new features for the following:

    – Pre-submission meeting,

    – Post-trial access

    – Waiver of local CT for new drugs already approved in certain other countries,

    – Accelerated approval under certain conditions, etc. and its preparation

A phase IV clinical trial is conducted with post-marketed products of new drugs under the following cases;

• In certain cases of new drug approval, the applicant is asked to conduct Phase IV clinical trial as per protocol approved by the local authority (LA), depending on the drug and disease for which it is indicated.

• Such conditions generally include drugs for serious/ life-threatening diseases for unmet need, rare diseases or diseases, which have special relevance to Indian health scenario, etc.

• In such cases, new drugs are approved with conditions requiring a phase IV clinical trial (CT).

For permission of clinical trials, the following documents have to be submitted

1. **Form 44**
2. **Treasury Challan of INR 50,000 (for Phase-I)/25,000/-(for Phase-II/III clinical trials).**
3. **Source of bulk drugs/raw materials.**
4. **Chemical and pharmaceutical information including**
   - Information on active ingredients:
   - Chemical name and Structure—Empirical formula, Molecular weight.
   - Analytical Data: Elemental analysis, Mass spectrum, NMR spectra, IR spectra, UV spectra, polymorphic identification.
   - Stability Studies: Data supporting stability in the intended container closure system during the clinical trial.

   d. Data on Formulation:
   - Dosage form
   - Composition
   - Master manufacturing formula
   - Details of the formulation (including inactive ingredients)
   - In process quality control check
   - Finished product specification and method of analysis
   - Excipient compatibility study
   - Validation of the analytical method
   - Stability Studies: Data supporting stability in the intended container closure system for the duration of the clinical trial.

5. **Animal Pharmacology**
   - Summary
   - Specific pharmacological actions
   - General pharmacological actions
   - Follow-up and supplemental safety pharmacology studies
   - Pharmacokinetics: absorption, distribution, metabolism, excretion

6. **Animal Toxicology**
   - General aspects

- Systemic toxicity studies
- Male fertility study
- Female reproduction and developmental toxicity studies
- Local toxicity
- Allergenicity/hypersensitivity
- Genotoxicity
- Carcinogenicity

7. Human/Clinical pharmacology (Phase I)

- Summary
- Specific pharmacological effects
- General pharmacological effects
- Pharmacokinetics, absorption, distribution, metabolism, excretion
- Pharmacodynamic/early measurement of drug activity

8. Therapeutic exploratory trials (Phase II)

- Summary
- Study report

9. Therapeutic confirmatory trials (Phase III)

- Summary
- The individual study reports with listings of sites

10. Special studies

- Summary
- Bioavailability/Bioequivalence
- Other studies, e.g., geriatrics, pediatrics, pregnant or nursing women

11. Regulatory status in other countries

a. Countries where the drug is:

    (i)   Marketed

    (ii)   Approved

    (iii)   Approved as IND.

    (iv)   Withdrawn, if any, with reasons

b. Restrictions on use, if any, in countries where marketed/approved

12. Prescribing information (of the drug circulated in other countries, if any)

13. Application in Form 12 along with T-Challan of requisite fees (in case of import of investigational products)

Medical devices- The authority regulating medical devices is the Central Drug Standard Control Organization (CDSCO) in the Ministry of Health, which lays down rules, standards and approves import and Manufacturing of Drugs, Diagnostics, Devices, and Cosmetics. Currently, CDSCO will be the approving authority for the import, manufacture and sale of medical devices in India.

In India, at present, only notified medical devices are regulated as drugs under the Drugs and Cosmetics Act of 1940 and rules made thereunder in 1945.

1. (i) substances used for in vitro diagnosis and surgical dressings, surgical bandages, surgical staples, surgical sutures, ligatures, blood and blood component collection bags with or without anticoagulant covered under sub-clause (i);

2. (ii) substances including mechanical contraceptives (condoms, intrauterine devices, tubal rings), disinfectants, and insecticides notified under sub-clause (ii); and devices notified from time to time under sub-clause (iv), of clause (b) of section 3 of the Drugs and Cosmetics Act, 1940.

The medical device regulatory tree in India is as follows as the following:
Ministry of Health and Family Welfare

> Drug Controller General of India - DCG (I)

> Central Drugs Standard Control Organization (CDSCO - Medical Devices Division)

The Ministry of Health and Family Welfare under Gazette notification declared the following sterile devices to be considered drugs:

1. Cardiac Stents

2. Drug-Eluting Stent

3. Catheter

4. Intra Ocular Lenses

5. I.V. Cannulae

6. Bone cement

7. Heart Valves

8. Scalp Vein Set

9. Orthopedic Implant

10. Internal Prosthetic Replacements.

***Pharmacovigilance*:** Pharmacovigilance is the best tool to establish the safety of a drug product. Every drug has some beneficial and undesirable or adverse effects. ADR are common clinical problems, which need minimization to decrease the economic burden of the country. In this context the launch of the Pharmacovigilance Programme of India (PvPI) marks an essential milestone toward safeguarding public health. PvPI is one of the indispensable steps taken by the Ministry of Health and Family Welfare, Government of India for ensuring safe and rational use of medicines. The program is coordinated by the Indian Pharmacopeia Commission, Ghaziabad, as a National Coordinating Centre (NCC). The center operates under the supervision of a Steering Committee (CDSCO).

Goal: Ensure that the benefits of the use of medicine outweigh the risks and thus safeguard the health of the Indian population.

Objectives:

- Monitor Adverse Drug Reactions (ADRs) in the Indian population.
- Create awareness among healthcare professionals about the importance of ADR reporting in India.
- Monitor benefit-risk profile of medicines.
- Generate independent, evidence-based recommendations on the safety of medicines.
- Support the CDSCO in formulating safety related regulatory decisions for medicines.
- Communicate the findings with all key stakeholders.
- Create a national center of excellence on a par with global drug safety monitoring standards.

Pharmacovigilance program of India (PvPI) has been started by CDSCO at the Indian Pharmacopeia Commission (IPC) and the following actions have been taken.

- Adverse drug reactions were collected through 210 AMCs throughout the county.
- AMCs report the ADRs through VigiFlow
- In 2015 launched Android Mobile App and helpline No.18001803024 (Toll-free) for ADR reporting
- PvPI collaborated with various National Health Programmes like AEFI, RNTCP, NACO, and NVBDCP

- Steering Committee monitor and supervise the Program Working Group approve technical and policy issues
- Quality Review Panel, Core Training Panel, and Signal Review Panel for quality review, training, and assessment of ADRs for potential signals
- Started Active Surveillance for MDRTB drug Bedaquilline in 6 identified Centres across the country
- Issued 37 drug safety alerts through SMS to create awareness in HCPs.
- PvPI makes recommendations to CDSCO for safety-related issues.

The Pharmacovigilance Program of India is administered and monitored by the following two committees:

1. Steering Committee

2. Strategic Advisory Committee

1. Steering Committee

- Chairman
- Drugs Controller General (India), New Delhi, ex- officio
- **Member Secretary**
- **Officer-in-Charge (New Drugs), CDSCO, New Delhi, ex-officio**
- Members
    1. Scientific Director, Indian Pharmacopoeia Commission, Ghaziabad, ex-officio
    2. Head of Department, Pharmacology, AIIMS, ex-officio
    3. Nominee of Director General, ICMR, ex-officio
    4. Assistant Director General (Extended Programme of Immunization as representative of Directorate General Health Services
    5. Under Secretary (Drugs Control) as representative of The Ministry of Health and family Welfare
    6. Nominee of Vice Chancellor of Medical/Pharmacy University, ex-officio
    7. Nominee of the Medical Council of India, ex-officio
    8. Nominee of the Pharmacy Council of India, ex-officio

Technical support will be provided by the following committees:

1. Signal Review Panel

2. Core Training Panel

3. Quality Review Panel

Governance Structure

- ADR MONITORING CENTRES MCI

- Approved Medical Colleges and Hospitals

- Private Hospitals

- Public Health Programs

- Autonomous Institutes (ICMR etc.)

- Collaboration with World Health Organization-Uppsala Monitoring Centre (UMC).

The PvPI National Coordinating Centre has collaborated with the WHO Collaborating Centre, Uppsala Monitoring Centre, Sweden, to achieve its long-term objective of establishing a "Centre of Excellence" for Pharmacovigilance in India.

A. Multiple Choice Questions

1. CDSCO Head is known as
    a) Director-General of India
    b) Governor-General of India
    c) Drug controller of India
    d) Drug controller general of India

2. Which Form no is required for application to import of new drug manufacturing?
    a) Form no. 10　　　　　　b) Form no .44
    c) Form no. 10a　　　　　　d) Both a and c

3. Function of CDSCO is
    a) Providing license for manufacturing of drugs
    b) Publish pharmacopeia
    c) Controlled the import and distribution of drugs.
    d) All

4. DTAB works under the guidance of
    a) DCGI　　　　　　　　b) CDSCO
    c) Government of India　　d) DCC

5. NDA application form required for biological and new drug applications?

    a) 397

    b) 356H

    c) 367

    d) 365H

6. The definition of Quality Risk Management (QRM) has been mentioned in the ICH guideline

    a) Q8

    b) Q9

    c) Q10

    d) Q12

7. Pharmacovigilance is a part of the

    (a) ICH E1 guidelines

    (b) ICH E3 guidelines

    (c) ICH E4 guidelines

    (d) ICH E2(A-F) guidelines

8. For medical testing laboratory accreditation provide by which ISO guideline

    a) ISO_15189–2007

    b) ISO_17025–2005

    c) ISO_17043–2010

    d) All

9. CDSCO headquarters are located in

    a) New Delhi

    b) Kolkata

    c) Mumbai

    d) Chennai

10. The primary responsibilities of the Central Drug Control Authorities is to regulate the

    a) Approval of the New Drugs and coordination with state,

    b) Grant permission to conduct Clinical Trials in the country,

    c) Control over the quality of imported Drugs

    d) All

11. The drug testing laboratory Mumbai is mainly testing of

    a) Medical devices

    b) Drugs and cosmetics

    c) Contraceptive devices

    d) all

12. Central research institute (CRI) is located at

    a) Kolkata

    b) Kasauli

    c) Lucknow

    d) Delhi

13. CDRI located in

    a) Lucknow

    b) Kolkata

    c) Delhi

    d) Mumbai

14. Uppsala Monitoring Centre, is located at
    a) Sweden
    b) INDIA
    c) London
    d) China

15. Central drug laboratory for veterinary testing is located in
    a) Izatnagar
    b) Delhi
    c) Mumbai
    d) Kolkata

16. State licensing authority responsible
    a) Verification of application form
    b) Inspection of premises
    c) Grant of licenses
    d) All

17. Drug inspector is appointed under the section
    a) 21
    b) 22
    c) 23
    d) 24

18. Common technical documents for non-clinical study reports are included
    a) M1
    b) M2
    c) M3
    d) M4

19. A common technical document for which module is not the part of CTD is
    a) M1
    b) M2
    c) M3
    d) M4

20. For the manufacturing of new drugs in India form no is required for application
    a) Form no. 10
    b) Form no.10A
    c) Form no. 40
    d) Form no. 46

B. Short Questions

1. What is CDSCO? Write down the main functions of CDSCO.

2. Write about various Drug Regulatory agencies.

3. Write about the different Central Drugs Testing Laboratories available in India.

4. Explain the Central Drugs Laboratory and its function.

5. What is COPP and its importance?

6. Write a short note on State Licensing authorities.

7. What are the general requirements for submission of application for the issue of COPP?

C. Long Questions

1. Define CDOSCO and write in detail about the standard procedure of new drug approval in India.

2. What is CoPP? Give details about the import process of the drug in India.

3. What is the CTD? Explain the different modules in detail.

4. Explain the state license authority audits function.

5. Explain about the state licensing authority and their organization and its function.

6. What are the regulatory requirements and approval procedures for new drugs?

MCQs Answers

1	2	3	4	5	6	7	8	9	10	11	12	13	14	15	16	17	18	19	20
D	D	D	A	B	B	D	B	A	D	C	B	A	A	A	D	A	D	A	D

Index

H

I

L

M

N

O

P

Q

R

S

T

U

V

W

Z